Process Machinery –
Safety and Reliability

Process Machinery – Safety and Reliability

IMechE Guides for the Process Industries

Edited by

Eur Ing W Wong
CEng, FIMechE, FIMarE

Mechanical Engineering Publications Limited
London and Bury St Edmunds, UK.

First published 1997

Reprinted 2002

© W Wong

ISBN 1 86058 046 7

A CIP catalogue record for this book is available from the British Library.

Printed in Great Britain by
Antony Rowe Ltd, Chippenham, Wiltshire

CONTENTS

ACKNOWLEDGEMENTS

The idea for this guide was conceived during the course of a one-day seminar, *Maximising Rotating Reliability*, organised by the Mechanical Reliability Committee of the Process Industries Division of the IMechE, held on 8 December 1994 at Headquarters. A working party was thereby set up with the task of planning the layout of the guide and providing material for its contents. The composition of the working party, was made up of the following members:

J. Harris	University Of Manchester
K.G. Rayner	ICI Engineering Technology
J. Lewis	ICI Eutech
D.T. Parr	AEA Technology
E. Parry	AEA Technology
J. Worsley	AEA Technology
W. Wong	Bechtel Ltd
D. Holland	Bechtel Ltd

Thanks are due to all the companies of the individuals concerned for the free use of their facilities during the course of the work. Special thanks are due to R. Moss for valuable comments and to D. Heckle for proof reading.

The final proof copy was sent to a number of manufacturers. The comments received from Weir Pumps were invaluable, especially their plea to reduce the Fog Index. Where some of the legal English is difficult, examples of interpretation have been added for clarification. We are also indebted to them for providing the compliance check list and the example of its application on a fire water pump as supplied for a typical offshore project.

Some useful comments were received from the British Pump Manufacturers Association and use has been made of material from the British Chemical Engineering Contractors Association and the Engineering Equipment and Material Users Association. It was also confirmed from the electric motor manufacturers that their interpretation is that all motors will be supplied as incorporated machines. Last but not least, thanks are also due to J. Harris, the Chairman of the Mechanical Reliability Committee, for all his work and encouragement.

HOW TO USE THE GUIDE

This guide is intended for the process industries. It provides a general interpretation of the Machinery Directive and other related legislation. The underlying objective is to ensure a common approach to health and safety. The guide meets this by outlining a procedure for the assessment of machine safety and reliability.

Public concern on health and safety issues has resulted in political action. There is now an ever-increasing enactment of safety, health, and environmental legislation throughout the world. Experience has shown that the previously prescriptive regulations are inadequate. This has resulted in new legislation which is objective and, therefore, open-ended.

In the past, requirements for health and safety could be satisfied by adhering to a checklist of requirements. The need for objective safety requires two undertakings to be demonstrated: first, that there has been sufficient effort to identify all risks to health and safety; second, that all reasonable and practical measures have been taken for their reduction. This is in addition to meeting all prescribed regulations, such as the Health and Safety at Work Act in the UK and similar regulations elsewhere. These relate to fire precautions, the safeguarding of moving parts, the control of dangerous substances, and the like.

In the section on legal requirements, an overview of the European Community Directives which are relevant to process plant is provided. Some workable principles of interpretation are suggested. Interpretation and legislation enacting the European Commission Directives will differ in each Member State; the regulations applicable in the country of interest must be consulted. Those relevant to the

UK are cross-referenced in Section 9. All interpretations are only opinions, even those of government departments! Authoritative rulings on the correct interpretation can only be made by the courts, the ultimate authority being the European Court of Justice.

The methods by which risks to health and safety can be identified and quantified are given in the sections on hazard and reliability assessment. Sufficient detail is given for the reader to become aware of the concepts and procedures which are commonly in use. The information needed for their implementation is identified, and this will subsequently enable a data package to be prepared for assessment. If necessary, consultants can be used to carry out the assessment.

The risk to health and safety will vary depending on the type of fluid being handled and the consequences of failure. In the section on the control of hazards, a classification procedure which takes account of this variation is proposed. A verification procedure is also given as a means of providing a final audit to ensure that all risks have been recognised.

A reference list, a reading list and a list of videos for distance learning is given for further study.

The appendices contain a list of definitions and suggested data sheets for materials safety, and checking machinery for health and safety compliance. An index list from the technical file relating to a fire water pump for an offshore platform is also provided, together with some sample documentation showing the use of the compliance check list, the application of FMEA for hazard identification and actions taken as a result of a HAZOP review.

RELATED TITLES

A Practical Guide to the Machinery Directive
H P van Ekelenburg, P Hoogerkamp, and D A Brown
ISBN 0 85298 973 3

The Design and Operation of Safe and Profitable Process Plant
IMechE Conference Transactions
ISBN 0 85298 854 0

Pressure Systems: Operations and Risk Management
IMechE Conference Transactions 1995–4
ISBN 0 85298 951 2

Engineering System Safety
G J Terry
ISBN 0 85298 781 1

The Reliability of Mechanical Systems
J Davidson
ISBN 0 85298 881 8

1 SCOPE

To provide a guide to good practice for the assessment of machine system reliability and hazards so as to:

- enable reliability (as related to safety) to be improved;
- enable hazards to be eliminated or controlled in a cost-effective manner;
- satisfy legislation, relative to the enactment of European Commission Directives, by the provision of guidance on appropriate records and documentation;
- develop an adequate cost-effective maintenance strategy.

2 INTRODUCTION

There is a legal requirement to control the hazards associated with machinery throughout the lifetime of the equipment.

The law now requires the design, operation and maintenance of machinery to be subject to a special review. This review is to enable the requirements for safety to be identified and prescribed. Safe use of the equipment throughout its operating life is the desired objective.

The reliability of a system or component is measured by its expected failure rate. Reliability becomes important in situations where the associated failure modes result in danger to health and safety. Another aspect of reliability is its impact on lost production, but this is not within the scope of this guide.

In the process industries the loss of containment of fluids is of major concern. This is because in most cases it will pose a threat to health, safety and/or the environment. Users of machinery in those industries will find this guide particularly useful.

There are established codes and standards applicable to the supply and construction of machinery and plant for the power and process industries. There are also established practices in safety assessment and identification which are commonly used in the process industry. These have, in general, provided safe and reliable plant and machinery. But accidents still occur – most of them (it is believed) due to failure to understand application limits and maintenance requirements.

This gap in communication has been recognised in the current (1995) and recent UK legislation. The Machinery Directive and the Construction (Design and Management) Regulations, are good examples of such concern: both of these measures are concerned with the adequate management and systematic recording of provisions for safe and reliable plant operation.

This guide identifies the practical measures required for compliance with the regulations, i.e. those measures which will provide adequate assessment of plant machinery safety and reliability. Users are warned, however, that the regulations are subject to interpretation, which, until tested in court, can provide no clear code of practice.

Hazardous situations caused by the wrong interaction between static plant and machinery should be considered as covered by the Construction (Design and Management) Regulations. This means that it is the duty of the plant designer to ensure that the interfaces between the plant and the machines are kept safe. All the operating conditions that can reasonably be foreseen must be taken into account.

3 LEGAL REQUIREMENTS

INTRODUCTION

The objective of creating a single 'common' market in the European Economic Area (EEA) goes back to the EEC Treaty (the Treaty of Rome) which originally established the Community.

Despite the elimination of tariff and quota restrictions between Member States, the Common Market was not yet a reality in 1985. The free movement of goods was still partially impeded by technical barriers such as differing national product standards. The growth of a free and competitive market for services was blocked by a range of national restrictions.

Achieving the free movement of goods was the basis of creating a single European market. The ministers responsible for trade therefore agreed on a 'new approach' to solve this problem for business. These 'New Approach Directives' (European Community laws) require Member States to be responsible for ensuring the health and safety on their territory of their people and in particular, of workers – notably in relation to the risks arising from the use of machinery. They set out 'essential requirements' (e.g. for safety), written in general terms, which had to be met before products could be sold in the UK or anywhere else in the Community.

The Community also agreed a common commercial policy covering trade relations with non-EEA countries. In many areas (mainly trade in goods) the Community now has obligations under the General Agreement on Tariffs and Trade (GATT) – for example, not to discriminate against the trade of other members of GATT and not to increase trade barriers without giving matching concessions in return. The purpose of GATT is to provide an accepted framework for the orderly conduct of international trade and to encourage the progressive opening of world markets.

The implications of the Community's approach are made clear by the following statement:

> *'The single market will be of benefit to Community and non-Community countries alike by ensuring continuing economic growth. The internal market will not close in on itself – 1992 Europe will be a partner, not 'Fortress Europe'. The internal market will be a decisive factor contributing to greater liberalisation of trade on the basis of the GATT principles of reciprocal and mutually advantageous arrangements.'*

The New Directives are generally of two types, the first of which is aimed at a broad product range rather than for specific products. In the second type, the detailed technical specifications are replaced by 'essential requirements' which describe the objectives to be achieved. In addition they refer to 'transposed harmonised standards' as a means of compliance with the essential requirements. It is this latter series of Directives with which this guide is concerned; a number of them are relevant to process plant, the most important being the Machinery Directive.

3.1 Machinery Directive (Supply of Machinery Safety Regulations)

The Machinery Directive is one of these 'New Approach' Directives. It is implemented in the UK by the Supply of Machinery (Safety) Regulations 1992.

From 1st January 1995 all 'new machinery' supplied in the EEA must comply with the Machinery Directive

(89/392/EEC). Compliance must be indicated by applying the CE mark to each machine. Some machinery will require type-examination before it can receive the CE mark. Any products legitimately bearing the CE mark can be supplied without hindrance within the EEA.

The original Directive has been amended twice; the first amendment is incorporated in the Regulations, but the second (93/44/EEC, adopted in June 1993) required additional UK legislation, the Supply of Machinery (Safety) (Amendment) Regulations 1994, which appeared early in 1996.

The concept of 'new machinery' is unfortunately not straightforward, in terms of either what is new or what is machinery.

What is clear is that a new model, either a one-off or for series production, introduced from 1st January 1995 must comply with the Regulations. In addition to this, three other groups of machines are considered to be new, viz.:

a) second-hand machines imported from outside (and never previously used in) the EU;
b) newly supplied examples of machines which have been supplied previously;
c) some refurbished machinery where the performance is improved from its original level.

The last item requires some interpretation:

- Refurbished machinery to be used for a different purpose is considered new.
- Refurbished machinery to be used for the same purpose, and to the same specification, is not considered new, as this work is routine maintenance.
- Refurbished machinery with any sort of modification has to be reviewed (see para. 3.1.9).

The second amendment broadens the scope of the Directive to include some classes of safety components and machinery for lifting people. It also adds some new 'Essential Health and Safety Requirements' (or 'EHSRs').

The underlying principle of the EHSRs is the requirement for safety integration. This means identifying and assessing the risks posed by the machine and eliminating them by good design rather than tacking on a proliferation of guards and safety devices. This may not always be possible, but the designer will have to demonstrate that all reasonable and practical measures have been taken. The EHSRs deal with such things as instructions, controls, lighting and an extensive range of hazard groups such as mechanical, electrical and noise.

An important feature of the Regulations is that some machinery which is considered to pose 'special hazards' has to follow a different route to conformity assessment. The classes of machinery involved are listed in Schedule 4 of the Regulations.

3.1.1   Individual Machines

The basic definition of machinery in the Directive is made up of four elements and it is concluded that all these must be present for a product to constitute 'machinery'. They are:

a) an assembly of linked parts or components;
b) at least one of which moves;
c) with the appropriate actuators, control and power circuits;
d) for a specific application.

The intention of the Directive is to have a definition which agrees with the commonly understood concept of a machine, as a product ready for use, unless it is specifically excluded by Articles 1.3, 1.4 or 1.5 of the Directive.

3.1.2   Assemblies of Machines

The second paragraph of Article 1.2 of the Directive deals with assemblies of relevant machines which, in order to achieve the same end – i.e. a specific application of the kind envisaged in the first paragraph – are arranged and controlled so that they function as an integrated whole. This is a matter which has given rise to much debate and is of particular interest to the process industries. The question that arises is: what is the status of integrated installations like oil refineries and chemical plant?

It could be argued that an oil refinery is an assembly of linked parts, some of which move, with the appropriate actuators joined together for a specific application, i.e. an individual machine.

With regard to the definitions and exclusions in Article 1, it is considered that to be within the scope of the Directive the assemblies must consist of machines in their own right as defined in the Directive, arranged with each other and any other products necessary for them to function as an integrated, controlled operation, for specific applications. Examples of what would be in scope would be vehicle assembly lines or assemblies in bottling plants and paper mills.

Individual machines within the assembly, which meet the definition of machinery in the first paragraph of Article 1.2 of the Directive and which are able to function independently, should be CE-marked by their supplier, who is responsible for fulfilling all the requirements of the Directive in relation to that individual product. The assembler of the machines is responsible for:

- selecting suitable products to make up the assembly;
- putting together the assembly in such a way that it complies with the provisions of the Directive;
- fulfilling all the requirements of the Directive in relation to the assembly;
- CE-marking the assembled whole.

There are, however, two exclusions in Article 1.3 of the Directive which should be noted:

- steam boilers, tanks and pressure vessels;
- storage tanks and pipelines for petrol, diesel fuel, inflammable liquids and dangerous substances.

In view of the nature of these installations, it is concluded, taking these two exclusions into account, that neither was it the intention to bring such installations within the scope of the Directive nor is it the result that such installations have been brought within the scope of this Directive. This means that they should be considered as separate stand-alone items. For example, a Pressure Equipment Directive will be issued in 1996.

It is our opinion that products which are not incorporated into machinery within the Directive's definition cannot therefore be classified as machinery for incorporation (Article 4.2, Machinery). The examples which follow illustrate how these requirements can be interpreted:

i) A pump can be incorporated as part of another machine, such as a gas turbine. They are both machines and the documentation can be combined, i.e. incorporated.

ii) A pump cannot be incorporated as part of a boiler. A feed water pump supplied as part of a boiler package must have its own documentation and must be CE-marked in accordance with the Machinery Directive.

iii) A boiler is a boiler subject to its own regulations and cannot be incorporated as part of a feed water pump as defined by the Machinery Directive.

iv) A compressor package can consist of a compressor, various pressure vessels, pumps and electric motors. While the pumps and the electric motors can be incorporated as part of the compressor, the pressure vessels cannot. Conversely, the compressor, pumps and electric motors cannot be incorporated with the pressure vessels. The packager, however, under the Construction (Design and Management) regulations, if not the Machinery Regulations, will be held responsible for the safety of the whole system.

1.3   *Machinery for Incorporation*

It is considered that 'machinery for incorporation' should satisfy the following requirements:

i)   It should meet the definition of machinery in Article 1.2, including being joined together for a specific application;
ii) While it can function independently, it should not be envisaged by the supplier as ready for end use, because it is designed solely to be incorporated into, or assembled with, other products which, taken as a whole, will also meet the definition of machinery in the first or second paragraphs of Article 1.2 of the Directive;
iii) The responsible person may, at his option, make a *declaration of incorporation*. The machinery is not then CE-marked until it has been incorporated into or assembled with other machinery. The declaration of incorporation is required to state specifically that the equipment to which it refers may not be put into service until the assembly has been declared to be in conformity with the Machinery Directive.

To interpret (ii) above, an electric motor can, in one sense, function independently: it will run if supplied with electricity. In the sense intended, it is envisaged by the supplier that it will be coupled to a load, such as a pump. The electric motor should therefore be incorporated with the pump as an assembled machine. By inference it then falls on the machine assembler to be responsible for compiling the technical file. There is no clear requirement that the supplier of the machine for incorporation should be obliged to compile a technical file for incorporation. It is suggested that the machine assembler can overcome this uncertainty by making such compilation a condition of purchase.

3.1.4   *The Technical File*

Objective of the technical file

a) In some directives the technical file is the principal means of assessment of the conformity of a product within the framework of the market surveillance by the EU Member States. In these cases, the assessment of conformity is based almost exclusively on the manufacturer's declaration of conformity, without the intervention of a third party or a notified body.

b) The file compiled by the manufacturer is intended essentially for the national inspection authorities' who have the right to ask the manufacturers or the importer to communicate the data relating to the tests carried out concerning safety, etc., when they have good grounds for believing that a product does not offer the degree of safety required in all respects. Refusal on the part of the manufacturer or the importer to communicate these data constitutes sufficient reason to doubt the presumption of conformity.

c) It must, therefore, be possible to place this technical file at the disposal of the competent national authorities should they so request when the product is placed on the Community market.

d) In other Directives, the documentation or technical file is just one of the means of completing a specific conformity assessment procedure with the intervention of a third party (notified body). This is the case with the Directives which provide only for the EC-type examination. This certificate is in turn included in the technical file.

This guide is concerned with the technical file referred to in paragraph (a) – that is to say, as the principal means of market surveillance without intervention by a notified body.

1.5 *Format and Contents of the Technical File*

In Decision 90/683/EEC of 13 December 1990 the Council established that 'the essential objective of a conformity assessment procedure is to enable the public authorities to ensure that products placed on the market conform to the requirements as expressed in the provisions of the Directives, in particular with regard to the health and safety of users and consumers.'

This guideline from the Council is, therefore, the essential criterion to be taken into account when considering the content and extent of the information to be supplied in the technical file provided for in the Directives, i.e. the content and extent of the obligation to provide information.

Consequently, the details included in the technical file always depend on the nature of the product and on what is necessary, from the technical point of view, to demonstrate the conformity of the product, either to the harmonised standards – if the manufacturer has followed them – or to the essential requirements of the relevant directive if the manufacturer has followed none, or only some, of the harmonised standards. This must, therefore, be determined case by case, depending on the product.

To allow effective exploitation of this file for market surveillance purposes, excessive paperwork should be avoided. To achieve this and to facilitate the manufacturer's task, it is proposed that the inspection authorities should accept sub-division of the file into two parts.

The first part (A) would consist of a summary of the essential technical data relevant to the conformity assessment procedures, including in particular:

- the name and the address of the manufacturer and the identification of the product;
- the list of harmonised standards followed by the manufacturer and/or the solutions adopted to satisfy the essential requirements;
- a description of the product;

- the operating instructions, if any;
- the overall plan of the product, if any.

The second part (B) would consist of a full file containing all the test reports, information concerning the quality manual, plans, descriptions of the products and processes, standards applied, etc.

If the manufacturer fails to follow this two-part breakdown of the technical file, the inspection authorities could demand the full technical file or part thereof, according to the requirements for inspection purposes, unless the details given in the declaration of conformity or in the certificate of conformity appear sufficient for the purposes of conducting a preliminary inspection.

3.1.6   Availability of the Technical File

The technical file must be kept at the disposal of the national authorities for inspection and control purposes. With certain exceptions, this obligation to keep at least one technical file inside the territory of the Community starts at the time of the placing of the product on the Community market, whatever the geographical origin of the product. This obligation is incumbent upon the manufacturer or his representative established in the Community. If the manufacturer is not established in the Community and has no representative in the Community, the person who places the product on the Community market must take on this obligation.

Any person responsible for placing a product on the Community market, but not in possession of the technical file, must be capable of:

- stating where the technical file is situated inside the Community;
- presenting the technical file as soon as possible on request from the national authorities.

However, the name and address of the person in possession of the file need not be expressly mentioned on the product or on its packaging, unless otherwise specified.

The file can be requested only during checks made for market surveillance purposes by the Member States. In any event, the request for the technical file must remain in proportion to the requirements of the inspection carried out. Therefore, in general, the manufacturer or person responsible for placing a product on the Community market should initially provide the inspection authorities with only a summary of the essential technical data (part A of the technical file). One or more specific points of the second part can nevertheless be requested in cases of serious doubt about the conformity of the product to the Community regulations.

The full technical file can only reasonably be requested where necessary, and certainly not when only an individual point is to be checked, in which case only the relevant part of the file should be required.

If the competent authorities in the Member State request the technical file, the first part of the technical file (part A) should be made available immediately, allowing a reasonable time for transmission. Extra time should be granted for submission of the second part (part B) of the file, taking into account its volume and form (written, computerised, etc.)

Community-wide organisation of the market surveillance procedures and co-ordination of the inspections should avoid repeated submission of the same technical file by the same manufacturer to different inspection authorities (cf. Sheet 11/E).

The technical file must be kept for at least ten years from the last date of manufacture of the product, unless the directive expressly provides for any other duration (cf. Council Decision of 13 December 1990).

3.1.7   *Language of the Technical File*

If the Community directives contain no specific provisions concerning the language of the file, the requirements of the Member States must be assessed on the basis of Article 30 of the EEC Treaty on a case-by-case basis, taking into consideration the proportionality of their demands. A Member State may request presentation of the first part of the technical file (part A) in its official language, but should not do so if the national authorities can understand the file or its contents in the other language. Where a translation is required, the person in possession of the file will be allowed extra time to submit the first part of the file to the inspection authorities. Moreover, no further conditions may be imposed concerning this translation, such as a requirement of a translator accredited or recognised by the public authorities, or of official translators or other similar requirements.

It should be noted that United Kingdom Regulations require that if the file is drawn up in the UK then English is to be used.

3.1.8   *Confidentiality*

Decision 90/683/EEC (Annex 1.1) stresses the need to ensure the legal protection of confidential information. No exceptions can be made to this very important principle, which the Member States must observe strictly. To this end, Member States must ensure that everyone involved in the assessments, inspections and surveillance who has knowledge of the contents of the technical file is bound to professional secrecy. Precise rules will, where necessary, have to be laid down by the Member States to guarantee this confidentiality. This applies in particular to the bodies notified by the Member States, which must ensure that these bodies maintain this confidentiality.

Confidentiality is also mentioned in the EN 45000 series of standards which serve as the reference standards for the notification of bodies by the Member States.

1.9   *Machinery Modifications*

The Machinery Directive requires that any modification of a machine requires the same procedure as for a new machine. In addition to design alterations, some changes in manufacturing arrangements and documentation should also be dealt with as modifications. Examples of this type include:

- significant changes to the quality system;
- revision of the hazard/risk assessment or other major sections of the technical file;
- for 'authorised representatives', changes in the relationship with the manufacturer.

Modifications within the requirements of the Machinery Directive are those which have an effect on safety. There is no simple definition of those modifications which should be regarded as safety-related, but in many instances this will be self-evident. The Machinery Directive implicitly requires a hazard/risk assessment of changes to a machine; carrying out this assessment should determine whether or not the modification has any potential safety significance.

It follows that an assessment must be carried for any modification, however small (see para 7.4).

If the assessment shows no impact on safety then a record of the assessment and technical details of the modification should be filed. For example, a pump with material changes, such as a shaft in some new material, could very well have no impact on safety.

If there is an impact on safety, then:

- an old machine with no CE mark will require a technical file in accordance with the Machinery Directive and a CE mark must be applied.
- Any existing CE mark on a machine will no longer be valid and a new one must be applied. The technical file must be updated in accordance with the Machinery Directive.

'Like for like' changes will usually have no impact on safety. Where changes of technology are involved, there

17

usually is such an impact. Changing a pneumatic control system to an electronic one could be said to fulfil the same function, but the failure modes will be different and safety could be affected.

It has been recognised that certain types of machines already meet the intent of the Directive by adherence to well-established and internationally recognised codes and standards. Such machines have specifically been excluded from the coverage of the Directive, because they are dealt with under separate legislation. The more important of them as regards process plant are as follows:

- steam boilers, tanks, and pressure vessels;
- storage tanks and pipelines for petrol, diesel fuel, inflammable liquids and dangerous substances.

Machines can be supplied ready for use, such as a fork lift truck, or as a component – such as an electric motor-driven pump – which may form part of an assembly. Each will need a CE mark. The excluded machines listed above may in themselves be composed of some machinery elements which require a CE mark. It can be seen that the term *machinery* has a very broad definition.

CE marks can be applied voluntarily by the manufacturer, when not subject to regulation and verification by an authorised agency. Verification may include the need for a type test.

The above is a summary of the more important points and is not a substitute for a study of the Directive itself.

3.1.10 Summary of actions required by the Regulations and Amendments

The supplier of 'relevant machinery' must ensure that:

1. the Essential Health and Safety Requirements as listed in Schedule 3 to the Regulations and as Amended are satisfied;
2. a conformity assessment has been carried out by the responsible person;

3. the responsible person has issued a declaration of 'conformity' or of 'incorporation', as appropriate;
4. a CE mark has been affixed, unless a 'declaration of incorporation' has been issued;
5. the relevant machinery is in fact safe;
6. a technical file has been drawn up and retained for ten years.

Any modification requires a review of all the work previously carried out. Any impact on safety requires a new or revised technical file to be prepared as appropriate. A new CE mark must be applied.

1.11 Machinery for Process Plant

It is accepted for the purposes of the Machinery Directive that the meaning of the words 'machinery and safety components' can have a wide or a narrow interpretation, and that this results in uncertainty regarding the scope of their application.

It was intended that the Directive would provide a clear and simple definition that suppliers could understand and could apply for themselves. However, individual Member States and the European Commission sometimes make different interpretations under the definition.

Some workable principles of interpretation, reflecting the substantive provisions of the Directive and its underlying intentions as revealed in the recitals, have therefore been suggested. It should be noted, however, that interpretations of these principles can only be statements of opinion; authoritative rulings on the correct interpretation of the provisions of the Directive and its implementing legislation can be made only by the Courts, ultimate authority in this respect being vested in the European Court of Justice.

Nevertheless, the interpretation given here should provide, in practice, a measure of confidence and the aim therefore is to provide process plant machinery purchasers and designers with some clarification about applying the requirements of the Machinery Directive to their plant.

3.1.12   *Process Plants and the Machinery Directive*

All process plants include some equipment which moves. A complete process plant – an oil refinery for example – could be said to fall within the definition of machinery quoted above. If this expansive view of the definition were adopted, it would lead to conclusions which would undermine the purpose of the Machinery Directive.

Specifically, every single item of moving equipment – every pump, compressor, control valve, etc. – could be supplied with a declaration of incorporation and not be CE-marked, on the grounds that it cannot function independently and that it is the complete process plant that should be declared in conformity with the essential requirements and CE-marked. This makes no sense, because it is the component equipment and not the complete process plant which is the subject of cross-border trade. Furthermore, the process plant may have only a relatively few items of *moving* equipment and a predominance of *fixed* equipment such as pressure vessels, tanks, pipelines, etc., which are expressly excluded from the scope of the Regulations. It is therefore difficult to see how CE-marking a complete process plant would be meaningful. Furthermore, both the Health and Safety at Work Act (HAWS) and the Provision and Use of Work Equipment Regulation (PUWER) impose obligations on the designer and operator to ensure that equipment of all kinds is safe.

3.1.13   *A Suggested Definition of Machinery for Process Plants*

It follows from the above that the definition of machinery should be interpreted to mean any assembly of equipment items which includes at least one moving item and which is necessarily connected together in order to process, treat, move or package a material, plus any control system needed to ensure that the assembly can be operated safely.

This would then exclude the fixed equipment which might be connected to this machinery but which is incidental to, or not necessary for, the safe operation of the machinery.

This would still leave contractors and users liable to comply with the Regulations and under the obligation to CE-mark assemblies of moving equipment with their associated system controls – where these assemblies are not being purchased in their entirety from one supplier.

An example would be a solids-handling system comprising a line of conveyors and elevators with an overall control system, safety switches, etc. Another would be a compressor driven by a gas turbine, where the contractor is purchasing the turbine and compressor separately and is designing the control system.

Although the declaration of incorporation does not refer to the technical file (except for Schedule 4 hazardous machinery), Regulation 24(1) requires the manufacturer or supplier to retain such a file for ten years. This file will normally contain design details and calculations which are not checked by the contractor or ultimate user, and it would be costly and unnecessary for the contractor (or user) to maintain duplicates of it. The contractor and user should therefore rely on the supplier to comply with this obligation – although it would be desirable to reinforce this by contract.

1.14 Interpretation of the Legal Requirements for a CE Mark

a) Users and contractors should limit the scope of the definition of machinery to the **Suggested Definition of Machinery** as given above.

b) Purchase orders for machinery or complete assemblies of machinery should require the supplier to issue a declaration of conformity and to CE-mark the complete assembly.

c) Where the user or contractor is the assembler of the machinery, as defined in paragraph 3.1.13 above, he should ensure that he is conforming with the essential requirements, is compiling and retaining a technical file,

21

is issuing a declaration of conformity and is affixing a CE mark. Exceptionally, he should ensure that he is meeting the special requirements for Schedule 4 hazardous machinery.

d) Where the user or contractor is purchasing machinery for incorporation into an assembly for which he will be taking overall responsibility, he should ensure that the suppliers of the components issue a declaration of incorporation and acknowledge that they are retaining the technical file required by the Regulations and that it will be made available on request by the user or contractor or by the HSE (or the equivalent enforcement authority in other Member States of the EU).

e) Purchase orders should explicitly state the user's or contractor's requirements regarding recommendations (a) – (d) above.

3.2   Other Directives

In addition to the Machinery Directive there are three other relevant Directives that are likely to be applicable to *machines which are incorporated into other machinery,* i.e. elements of a machinery package.

It should be noted that the scope of a Directive is limited to the risks covered – that is, where the risks referred to in the Machinery Directive are wholly or partly covered by other specific Directives the Machinery Directive shall not apply.

3.2.1   *Electromagnetic Compatibility (EMC) Directive*

This Directive (implemented by the EMC Regulations) requires that 'apparatus shall be so constructed that:

- the electromagnetic disturbance it generates does not exceed a level allowing radio and telecommunications equipment and other relevant apparatus to operate as intended; and

- it has a level of intrinsic immunity which is adequate to enable it to operate as intended when it is properly installed and maintained, and used for the purpose intended.'

As an example, a software process control system must not be capable of being affected to the extent of preventing it from operating as intended as a result of electromagnetic interference from (say) a fluorescent light; neither must it cause any equipment to be affected by the emission of electromagnetic radiation.

The date of enforcement of this legislation is January 1 1996. For the EMC there are three routes to compliance:

i) The **standards route**: this is available for apparatus other than radio communication transmitting apparatus. It involves self-certification by the manufacturer against the appropriate harmonised European standards.

ii) The **technical construction file route**: this is available for apparatus other than radio communication transmitting apparatus, where there are no European harmonised standards or where the manufacturer chooses not to apply such standards. The technical construction file includes a report or certificate issued by a 'competent body' appointed by the DTI.

iii) The **EC type-examination route** for radio communication transmitting apparatus, including mobile telephones. An EC type-examination certificate is issued by a 'notified body' appointed by the DTI.

2.2   *Low Voltage Directive (LVD)*

This Directive, implemented by The Electrical Equipment (Safety) Regulations 1994, requires that 'electrical equipment must comply with the requirements of regulation 5(i).' That is to say, it must:

- be safe; and
- be constructed in accordance with principles generally accepted within Member States as constituting good engineering practice in relation to safety matters; and

- conform to the Principle Elements of the Safety Objectives of Schedule 3 of the regulations.

The date of enforcement of this legislation is January 1 1997. For the Low Voltage Directive, the route to compliance is via the technical file route.

3.2.3 *Pressure Equipment Directive (PED) and Simple Pressure Vessels Directive (SPVD)*

The above directives are two other New Approach Directives based on product safety. Many more are expected to follow. These directives are broadly similar to the other Directives given in more detail above. The products specifically have to:

- meet essential safety requirements;
- have safety clearance from an approved body;
- bear the CE mark;
- be provided with technical documentation;
- be accompanied by manufacturers information;
- be *safe*.

3.3 Construction (Design and Management) Regulations (CDM Regulations)

The CDM Regulations place duties on all those who can contribute to the health and safety of a construction project. Duties are placed upon clients, designers and contractors, and the Regulations create a new duty holder – the planning supervisor. They also introduce new documents – health and safety plans and the health and safety file.

The Regulations cover the installation, commissioning, maintenance, repair or removal of mechanical, electrical, gas, compressed air, hydraulic, telecommunication, computer, or similar services which are normally part of a structure. A structure is defined to include fixed manufacturing plant which involves construction work over 2 metres in height (i.e. process plant).

.3.1 The Client (on initiation of a project)

The client is required to:

- ensure that financial provision is made and time is allowed for safety requirements in the initial planning of a project;
- ensure that prior notice is given to the nominated authority (the Health and Safety Executive in the UK) of the project if it is expected to last more than 30 days or involve more than 500 people;
- establish the site development requirements, identify any applicable hazards, produce the project conceptual design and issue it to the project team;
- appoint a planning supervisor from the project design team;
- appoint a competent contractor as principal contractor.

.3.2 The Planning Supervisor

The Planning Supervisor's duty is in effect to ensure that all safety regulations have been recognised and have been compiled with at all stages of the project. The major outcome of this work is the production of the Safety Plan and the Safety File.

The Safety Plan

The plan should show how it is proposed that all hazards are to be identified and that risks to health and safety are to be lowered to an acceptable level throughout all stages of the project. It will include a list of scheduled activities and procedures to be used from design through to construction and handover. Typical for process plants will be:

- the setting-up of the safety team and the assignment of duties;
- identification of hazardous materials in construction and operation;

- producing a list of documents for safety review;
- identification of required safety drawings;
- ensuring that required safety equipment and facilities are provided;
- scheduling of HAZOP and HAZAN meetings;
- classifying and identifying critical machines and processes;
- nominating machines for hazard assessment;
- verifying adequate maintenance facilities for safe access;
- co-ordinating safety and constructability meetings between design and construction;
- ensuring that safety-centered maintenance plans are prepared;
- monitoring provision for safety activities with regard to time and cost;
- ensuring adequate training and safety management provisions for construction;
- maintaining records and preparing the safety file.

The Safety File

The safety file is required to contain a record of all the 'as built' design features, including all the information on risks to health and safety that could arise from operations and maintenance, and the maintenance tasks needed for safe operation.

The file must be given to construction for updating and finally checked by the planning supervisor before being formally handed over to the client prior to operation.

3.3.3 *The Designer*

The designer is required to identify any risks to health and safety in the design which could arise during construction, operation or maintenance, either from the materials used or the facilities provided. The design in question must include all reasonable and practical features to avoid these risks in accordance with the principle of safety integration. The designer must:

26

- make clients aware of their duties under the Regulations;
- give due regard in the course of his/her work to health and safety;
- provide adequate information, to those who need it, about the risks to health and safety of the design;
- co-operate with the planning supervisor and, where appropriate, other designers involved in the project.

Design is taken to mean all necessary drawings and documentation.

3.4 *The Client (as operator)*

The client has to receive the safety file prior to handover of the completed project and is responsible for its safe keeping for future reference. The client has a duty to consult the file concerning any maintenance work or any subsequent alterations to the plant. This in effect overlaps with the PUWER Regulations which follow.

3.4 Provision and Use of Work Equipment Regulations (PUWER) 1992

These regulations lay down important health and safety laws for the provision and use of work equipment. Information on these regulations is covered in HSE Document L22, *Guidance on Regulations*.

The term 'work equipment' means any machinery, appliance, apparatus or tool and any assembly of components which, in order to achieve a common end, are arranged and controlled to function as a whole. The requirements imposed on an employer by this legislation apply to work equipment provided for use, or used by, any of his employees.

The regulations come into force in two stages: a first group of regulations from 1st January 1993 and the remainder from 1st January 1997. The first group requires that the employer shall ensure that the work equipment is so constructed or adapted as to be suitable for the purpose

27

for which it is provided. The equipment shall be maintained in a safe state, the maintenance work being carried out by knowledgeable and trained people. These regulations also cover requirements for information and training, and the need for the users of equipment to comply with other relevant Community Directives.

The remainder of the regulations cover access to dangerous parts of machinery, protection against specific hazards, high or low temperature applications, control and control systems, isolation from sources of energy, stability, lighting, maintenance operations, markings and warnings.

Machinery provided in a process plant to be operated by workers as part of a production process can be considered as work equipment.

4 HAZARD ASSESSMENT

INTRODUCTION

Hazard Assessment is the identification of all possible sources of danger to safety and reliability. Some typical examples are given in Appendix B.

The danger could be inherent in the process, such as handling of a poisonous gas. It could be the result of a component failure. Very often it could result from operator error during transient conditions, such as starting, stopping, a change of throughput, or change of process specification. A change of state from operation to maintenance which requires isolation and purging of toxic gas can be dangerous if proper facilities are not provided.

The Machinery Directive requires that these dangers must be identified so that design measures can be taken for their avoidance.

.1 Identification of Hazards

The Machinery Directive requires a systematic review of a list of possible hazards (Ref. British Standard EN 414) in order to identify under what circumstances they would be applicable. This is usually carried out by a multidisciplinary team in accordance with procedures as given in the Chemical Industries Association booklet *'A Guide to Hazard and Operability Studies'*.

4.2 Machine Hazard Assessment

Care must be taken to ensure fitness for purpose of machinery and compliance with all relevant HSE regulations and guidance notes. For simple machines this may take the form of a checklist which can be completed by a competent person with a full understanding of safety and HSE requirements.

A reliability assessment may also form a critical part of a hazard assessment, particularly where the subject machine is in toxic, flammable or explosive service. The purpose of this reliability review is to ensure that historical data is used to optimise maintenance routines and to highlight possible failure modes and their consequential effects. The results of this exercise will then be used in specifying adequate provision of spares and also for determining the possible need for additional safety equipment.

The importance of the machinery as a part of a complete process train or unit must always be appreciated.

4.3 Information Required

The information required to carry out a hazard assessment of the machine covers the design intent of the machine, the actual process requirements, the operating controls on the machine, and the maintenance applied to the machine. Reviews of operating experience and the machine's containment capability will be included with the Design Verification Report covering the aspects of the machine studied to ensure satisfactory minimisation of the identified risks. The information should also include a listing of the technical codes and standards used in the design.

5 RELIABILITY ASSESSMENT

INTRODUCTION

Failure of equipment to perform its intended purpose in a reliable manner can result in unsafe scenarios. In order to evaluate reliability, in practice it is usual to express it in terms of its corollary, 'unreliability' and its associated 'failure'. In process plant machinery, the various modes of failure may be numerous, and the likelihood and consequences of a failure may be difficult to evaluate. The loss of production as a result of unreliable machinery is a commercial issue, but unreliability which results in a risk to health and safety could be a criminal offence. In order to comply with current legislation it is necessary to demonstrate that the machinery has been made risk-free *as far as reasonably practicable*. It is believed that the suitable application of formalised methods of analysis and assessment will serve to identify the *reasonable practicable* measures needed to reduce risk.

The following notes provide an introduction to the concepts involved.

5.1 Failure Rate (Failure Frequency)

The machine structure consists of machine elements and supporting systems which in turn have to be integrated into a process plant. Quantification of risk has to rely on statistical data. Failure rate is a measure of the frequency per unit of time that a machine or component breaks down. Data is available for most process machinery and system elements from published sources such as the OREDA offshore reliability data handbook. It is important to understand that this data is by its nature general and not specific. The data can be used to forecast the probability of events which may be severely affected by local circumstances. It is recommended that specific data is gathered from critical machinery, once in operation, for constant review. Failure rate is frequently assumed to be constant and not affected by component degradation.

5.2 Reliability Prediction

Reliability may be defined as the probability that a machine will perform a required function under stated conditions for a stated period of time. It is based on a function of the failure rate and the average time taken to reinstate the machine following a failure.

With mechanical systems, and particularly those containing rotating machinery, the prediction of reliability performance is not straightforward. There can be considerable variance in the failure frequencies and average repair times of components, and these must be taken into account. Furthermore:

- There is a need to consider and limit the effects of component degradation.
- Operational, environmental and maintenance conditions which may affect the validity of the generic failure rate data used in assessing equipment and system availability also need to be studied.

- A subjective evaluation of the factors which may identify any uncertainties in the basic data is also necessary.

These measures should ensure that predictions based on the assumption of the constant failure rates adopted are sound so that design decisions can be taken with a reasonable degree of confidence.

> *Take for example the reliability of a Gas Turbine, and its ability to maintain rated output. It is affected by the degradation of its combustion system, compressor, and turbine. These are affected by the adopted margin between required power and the rated power, any need for cyclic operation, frequent starting and stopping, and the site conditions such as atmospheric dirt and pollution. Furthermore there is also an effect from the type of inlet filtration, the type of compressor cleaning system, and the type of fuel used.*
>
> *Repair time is made up of the time needed to operate a permit system, making ready for maintenance work and the mobilization of tools and spare parts as well as the actual time taken to execute the repair. Overall repair time will depend on the efficiency of the staff involved and may vary considerably from site to site.*
>
> *All the above considerations will affect reliability.*

Failure rate data is needed for the evaluation of mechanical reliability. The use of analytical techniques such as failure mode and effect analysis (FMEA) and fault tree analysis (FTA) is also involved.

3.2.1   *Failure Mode and Effect Analysis (FMEA)*

Failure mode and effect analysis is a step-by-step procedure for the systematic evaluation of the failure effects and the criticality of potential failure modes in plant and equipment. It has many different applications and can

33

be applied at different levels of detail called *indenture levels*. For example, it may be used to determine the likelihood of breakdown of a gas compressor or the probability that a fire protection system will fail to operate when required to do so. At a more detailed level, it could be employed to provide an evaluation of the failure mechanisms associated with a pressure sensor which could be an essential part of the fire protection system.

By analysing the failure modes of individual items the effect on machine and system operations can be identified. The need to take action and what measures should be taken can be judged by the criticality of the effect identified.

FMEA is a 'Bottom up' approach. Each potential failure mode in a system is analysed to determine its effect on system reliability and performance. The effects can then be classified according to severity. All contributing factors (to loss of safety) need to be drawn out with all the relevant engineers and specialists present, in order to arrive at an effective solution in a cost effective manner.

The methodology is based on the use of a spread sheet and the use of key headings. It relies on the expertise of the participants and their skills in determining the possible failure modes.

5.2.2   *Fault Tree Analysis (FTA)*

FTA is a 'Top Down' approach. Starting from the definition of a top unwanted event, such as 'Loss of Machine Control', all possible paths to failure are traced, down to the basic set of initiating events. These basic events are combined by AND/OR gates into a logic tree.

An analysis of the logic tree requires the use of Boolean Algebra so that an expression can be found for the top unwanted event. Generic failure rates can be applied for each element together with estimated repair times and required inspection intervals to calculate the failure probability. From the fault tree the main contributing factors can be identified and measures to reduce the failure rate explored.

.3 Factors Affecting Reliability Assessment

The operating regime, environmental conditions and maintenance strategy, as already mentioned, are the factors that need particular attention in mechanical system reliability assessments. Failure mode and effect analysis, in association with fault trees and a subjective analysis of equipment maintainability, should ensure that the potential critical failure modes associated with material degradation, such as with seals and packing, are identified at the design stage.

Appropriate condition monitoring systems should be specified to ensure that these potentially critical failures are revealed before components deteriorate significantly. Maintenance actions can then be planned for the next time the plant requires a shutdown.

Planned maintenance can be used based on predicted availability if degradation type failures are anticipated, and the eventual corrective action can then be pre-planned. This will then be the best that can be achieved without introducing additional redundancy or other design changes.

However, the failure tolerance of alternative systems, such as the use of glandless pumps instead of an existing pump with mechanical seals, can in some cases be significantly higher and the cost of replacement pumps may be a worthwhile price to pay for a system with a higher availability potential.

.4 Probability of Failure

Failure of a machine in continuous operation is immediately apparent. As with the measure of reliability, the probability of failure can be defined as the probable fractional time during which the machine or system does not function as intended. The reliability of such systems can be improved by increased redundancy, such as the installation of stand-by machinery.

5.5   Risk

Risk can be defined as the probability of a specified unwanted event. In the case of electricity supply for example, the risk will be the probable period of time that there will be a loss of supply during the period of time under consideration.

5.6   Probability of Failure on Demand

In certain operational situations machinery system failure may not become evident until a demand or a test is made on a system - for example a Fire Water pump. Its failure is not known until it is required to be used. In order to improve its chances of being available it is required to be tested on a regular basis in order to detect any fault so that it can be repaired. The probability of failure on demand is then a function of the equipment 'unrevealed' failure rate and the test interval. The reliability of such systems can be improved by increasing the frequency of testing.

In this situation the risk will be that the Fire Water pump will not function when there is a fire. This will need to be based on the probability of a fire and the fire pump not working. This type of analysis is required for all types of safety systems such fire and gas detection and ESD (emergency shut down).

5.7   The Human Interface

Plant and machinery require human intervention, however small. Telltale trends, signals and alarms can give warning of incipient machine failure. Operator skill in interpreting the signs and the ability to take speedy action to avert disaster will affect the consequences of machinery failure. The location, accessibility, and ease of operation of any controls that need to be used in an emergency will affect the time of response.

In assessing the consequences of failure, the workforce characteristics must, therefore, be taken into account.

The demographic statistics of importance will be the level of intelligence, education, motivation and training.

The metrology of the workforce will be the statistics of its physical dimensions and condition. A plant or machine designed to be operated by trained athletes with university degrees cannot be safely operated by illiterates.

.8   Risk Assessment

A flow diagram illustrating the steps required before arriving at a risk assessment is shown in Figure 3.

This shows that risk can be defined as a product, of the consequential losses associated with an unwanted event, and its probability of its occurrence. Risk assessment is the act of weighing the balance between the chance of a failure occurring and the resulting consequence. A Risk exposure index can be employed to rank different events in order of importance so as to provide justification for spending money to avoid, or to mitigate the effect of the risk.

A model for risk assessment is given in Figure 4.

.9   Reliability Specification

It is recommended that for all large or complex process plant machinery, a reliability specification is included in the purchasing specification if the likelihood and consequences of failure will have a major impact. Model clauses which may represent a reliability specification are detailed later.

The object of such a specification is to ensure that consideration is given to reliability and availability in order that:

1) a minimum requirement is achieved;
2) opportunities to optimise the design for reliability and availability are not overlooked.

Wherever possible, reliability targets should be set, in order:

1) to match, or better, those of similar machinery;
2) to meet purchaser's needs in relation to a combination of risk, cost and benefit.

It is recognised that it may not always be possible to set formal targets, and in this case the purchasing specifications should still require reliability/availability assessments to be executed and the machinery design reviewed for possible improvements. Cost-benefit analysis should then be applied to enable sound judgments to be made.

As part of the purchasing cycle, reliability specifications should be included in the inquiry documents to ensure that the selection considers reliability/ availability. The potential supplier, even if he does not have information to hand on reliability, should be required to state in his tender how he intends to address this topic in the event of an order. It may require additional payment which should be identified and agreed before order placement.

It is stressed that such assessments should not cause major or insurmountable problems to any manufacturer or supplier. By referring to BS5760, published works or consultants, a manufacturer or supplier should be able to use these techniques to facilitate the conclusion of a satisfactory order. Historical data on similar machines may at first appear to be scarce, but data banks do exist in the public domain and obviously machines are constructed of many simple parts for which data are available.

5.10   Model Safety and Reliability Specification

It is recommended that the following model clauses (as published by the Engineering Equipment and Material Users' Association, EEMUA) should be included in purchasing specifications for all major items or packages of capital equipment, particularly in any case where the extent of supply includes anything that could be considered a

38

'system' (for example, the lubrication system supplied with a rotating machine).

Paying specific attention to reliability issues which affect health and safety in effect reinforces the requirements of the Machinery Directive.

1) The vendor shall demonstrate that he has considered the reliability/availability aspects of the package he is proposing to supply in a disciplined and structured manner. The object of this consideration shall be:

   i. to identify critical failure modes, especially those that have an impact on health and safety;

   ii. to estimate the overall reliability/availability of the total package in relation to safety issues and the duty or duties which it is intended to perform;

   iii. to identify those components of the package which make a major contribution to unreliability or unavailability and risk to health and safety;

   iv. to provide confidence that major potential weaknesses in the package have been estimated.

2) An acceptable procedure which would satisfy the above requirements would comprise:

   i. construction of a reliability block diagram where each 'block' represents an item, component or sub-system for which meaningful reliability data can be provided;

   ii. identification of the potential failure modes of the 'blocks' and the consequences of such failures on the performance of the package;

   iii. by using a fault tree, or an equivalent alternative method, computation of the overall unreliability/unavailability of the package and identification of significant contributors;

   iv. proposal of modifications for consideration by the purchaser to improve the design.

3) The vendor shall declare the data used, the sources thereof and any assumptions made. The purchaser may wish to review this data and discuss its applicability to the particular equipment. The vendor may be required to substantiate any data that is not within the public domain.

4) In making assessments of availability, the vendor will need to estimate repair times. For the purpose of making such estimates, the vendor may assume that the spares holding is in accordance with his submitted recommendations unless otherwise stated. He may also neglect any delay caused by factors not directly related to the package in question, unless a basis for estimating the extent of such delays is provided by the purchaser. In the case of failure modes which make a significant contribution to the overall unavailability the vendor may be required to demonstrate that his assumed repair times can be achieved.

5) The overall availability/reliability of the whole package, as predicted by the analysis shall not be less than the value stated in the inquiry or order or such other value as may be agreed.

6) The vendor's warranty for the package shall not end until such time that the reliability/availability achieved in an agreed trial period has been demonstrated to be not less than that predicted, or some other value as may be agreed. The length of the trial period shall not be less than ten times the predicted overall mean time between failures, or some other period as may be agreed, but in no case less than six months.

Comments on the above specification
Difficulties will usually be experienced in obtaining guarantees and/or warranties with respect to machinery

reliability. Detailed negotiations should be expected with particular reference to the issues as given in paragraph 5.2 above. In some cases, to overcome these problems so as to obtain a guarantee, it may be possible to place a contract for maintenance, or in some cases total operations and maintenance, with the vendor. As a minimum, however, vendors can be required to guarantee and demonstrate repair times as suggested in paragraph 4 of the above specification.

CONTROL OF HAZARDS

INTRODUCTION

The combined effect of the CDM and Machinery Directives results in the need to ensure that the needs of safety is taken into account from project conception through design, construction and into service. In the procurement of any machinery that may be required, each of the parties concerned have specific responsibilities to fulfil.

A procedure is proposed for the control of hazards by a system of machine classification during design and verification that the design is safe before operation.

1 Responsibilities

1.1 The Client

The concept of any project originates from the client, who specifies the duty and rate of output from the project, its location and the way in which it will be operated. This results in the specification of the required machinery. The client also has the ultimate authority in deciding the machinery vendor to be selected.

When the machinery is accepted into service, the client then has the responsibility for the operational and maintenance policy. He must ensure operation within design limits, control all hazards as identified to him and carry all necessary maintenance.

6.1.2 *The Designer*

The designer converts a concept into a detailed design, determining the detailed specification of the machine and all the operating conditions to which it will be exposed, both transient and steady-state. The designer also determines the type of machine required and may select suitable machinery vendors for consideration by the client.

a) Project Responsibilities

It is the responsibility of the design project manager to ensure that the appropriate machine system classification has been completed in accordance with the Machinery Directive. The design project manager is also responsible for the appointment of a safety planning supervisor in accordance with the CDM Regulations.

b) Safety Planning Supervisor

The safety planning supervisor is responsible for the health and safety plan and for the preparation of the safety file in accordance with CDM regulations.

c) Machine Design Responsibilities

For critical machine systems it is the responsibility of the design engineer to oversee the integration of the machine system into the process. This activity includes:

i) agreeing the technical specification of the machine system;

ii) choosing the vendor and specifying any inspection requirements;

iii) identifying hazards and failure modes;

iv) agreeing with the vendor on any protective systems and operational or maintenance requirements, to ensure safe operation and to meet legal requirements;

v) obtaining design verification from a specialist engineer;

vi)   specifying commissioning tests to ensure the safe functioning of the system.

d)   Safety or Loss-Prevention Engineer

The safety or loss-prevention engineer is responsible for carrying out the safety assessment and reviewing all identified hazards and failure modes in accordance with the health and safety plan as prepared by the safety planning supervisor. He monitors the safety plan, chairs all HAZOP meetings and safety audits, and prescribes steps to ameliorate risks to health and safety.

For small projects his work can be amalgamated with that of the safety planning supervisor. For large projects there may be a team of safety engineers.

1.3   *The Machine Vendor*

It is the responsibility of the vendor or his representative to design the machine system to meet the specified duty points and to provide a complete statement of design limits which will ensure machine safety relative to all identified failure modes.

The vendor shall be required to affix a CE Mark to the machine and issue a declaration of conformity in accordance with EEC Directive 89/392/EEC (UK Statutory Instrument SI 1073/94).

It should be noted that, if loss of machine containment is to be avoided, provisions additional to the equipment package may be needed, such as protective devices to alarm, trip or isolate. In such cases the plant designer and machine vendor will need to determine who holds the master technical file. This establishes responsibility for carrying out the work of incorporation and applying the CE stamp. In general, this should be the work of the vendor.

1.4   *The Construction Contractor*

It is the responsibility of the site construction manager to ensure that the safety plan has been studied and that the required precautions, procedures and training of construction workers are carried out. Construction input will also be required during the design phase to ensure that the requirements of the safety plan can be implemented. Refer to the CDM Regulations for the responsibilities and requirements.

6.1.5 The Construction/Commissioning Inspector

The commissioning engineer shall ensure that, prior to commissioning, the equipment and its installation are checked for compliance with the design intent .

He shall ensure that tests as specified under 6.1.2 (c) (vi) are carried out to prove the correct function of the system before hazardous materials are introduced to the plant.

6.1.6 The Site (Plant) Manager

The site or plant manager is in effect the client's or clients' nominee: it is his responsibility to verify that the plant and its machinery are fit for their intended purpose before accepting them into service. On handover the manager receives the safety file and any technical files for safe keeping and review. He then becomes responsible for the safe operation and maintenance of the plant and its machinery. See paragraph 6.1.1.

6.2 Classification of Machines

In this context, classification is a procedure for ranking machines with regard to risk. The potential hazards of the machines are considered and the machines are classified as either critical or non-critical. This judgement is based on assessment of the hazard presented by the process fluid and on the potential consequence of its release, the significance of process consequences following incorrect operation of the process or failure of the machine, and the potential damage due to mechanical failure given the nature of the machine assessed.

Significance of Classification

The classification process described achieves two objectives: firstly the prioritisation of the machines as to overall hazard, and secondly the establishment of criteria against which the verification requires to be conducted.

The detailed procedure for examining a machine is built around a series of reviews. In order to determine the focus of these reviews, i.e. to decide which sort of review to carry out, it is vital to understand and agree the basis for the criticality classification. This may require clarification of the original intention of an installation, where this is not apparent from the information that has been presented.

The classification of a pump as critical for reasons of process consequence required the verification engineer to have a clear understanding of what these 'process consequences' were. In the particular case the machine appeared to be adequately matched for the duty. However, the process intent was such that a failure of the pump would have unacceptable safety and reliability consequences. Though the pump could be expected to be reliable, pump failure could not be ruled out.

The design had not included any particular monitoring of process or machine parameters to give prediction of such failure, and had not included the installation of a spare or of a system to provide cover in the event of failure of the on-line unit. When this information was communicated to the project design team by the verification engineer, it was appreciated that the plant design had not implemented the original intent of the process engineer.

6.3   Machine Verification

The verification process looks at the machine and the process in which it operates. It therefore provides an independent review prior to the commissioning of key areas of the production process. This systematic review identifies points of concern where the original requirement is not met by the proposed equipment.

The classification of machines takes place alongside other design reviews at the start of the design process. The verification study provides a detailed check at the completion of design, manufacturing and installation of the equipment. It can ensure that the original intent for the equipment is met safely.

The verification review forces the design process to be complete prior to commissioning, and ensures that all design needs have been addressed.

Detail is important to successful machine operation. For example, incorrect package units may be supplied because materials were wrongly specified on the purchase order or on the manufacturer's standard. This will only be detected once the machine dossier is submitted. It should be detected by the design engineer, but this is not always the case because the engineer at the end of a project may not be the same one as at the start.

Consistency of intent is difficult to achieve where the equipment is called for as part of a front-end engineering package and is designed by the manufacturer from a design contractor's specification. The verification engineer needs to review the whole concept of the design, bringing the key safety points into the limelight and throwing deviations from the original intent into relief.

Verification Methodology

The verification analysis of the information provided must be structured so as to facilitate a logical and complete assessment. The structured information which will then be available can then be reviewed as a whole to provide a consistent picture of the capability of the machine system.

3.1 *Machine Purpose and Specification*

This information is required to determine the suitability of the machine for the intended task. This would ideally come from three sources:

a) The process data sheet, which outlines the scope of the process design.

b) The mechanical data sheet, where the process design is translated into machine requirements incorporating the safety and other best practice points for machine reliability relevant to the particular installation.

c) The machine supplier's data: the translation of the machine requirements by the supplier must be sufficiently detailed to show that the machine will be capable of meeting the full range of conditions (including start-up and shut-down) required by the process.

In the case of a new installation, all this information must be available and should be supported in particular instances by design reviews of the installation, such as material-of-construction reviews, relief-and-blow-down reviews, etc. This is illustrated by the following example.

In the case of an ammonia screw compressor design review, it was found that the logic of the pressure relief of the compression system, in accordance with normal procedure, was built around the capacity of the machine. However, screw compressors are designed with an internal compression ratio which is unaffected by any discharge pressure relief. It was found that high suction pressures could occur at start-up which could result in internal pressures exceeding the casing design pressure. Additional operating reviews and manufacturer reviews were needed before the safety of the installation could be demonstrated.

6.3.2 Machine Operation and Protection

As with the machine purpose and specification exercise, information on machine operation and protection can be found in several areas:

a) The engineering line diagram. This is an essential document because it fixes the position of the machine in a frame of reference for the process. Examination of this information has much in common with the techniques used for hazard studies, where normal and extreme conditions must be considered against different operating situations. Though the formal hazard study reports will be considered later in the review process (if available) the view of the Line Diagram from a 'machines' point of view can give a different aspect from that derived from the formal study – which tends to cover process streams rather than machine effects.

b) Instrument schedules. These documents provide a wealth of information in terms of confirmation that the protection identified on the line diagram is in place and that the actual limiting values associated with a protective device are correct. The reason for the device is sometimes obscure and it may be necessary to add notes on its function.

Cases have occurred where the protective device was available and in use; however, examination of the instrument maintenance schedule showed it to be untested. The reason for the protective device was not understood and the need for it to be regularly tested to ensure its reliability was therefore not recognised.

c) Hazard and operability reviews. These have been produced for projects over the last twenty years, and can be extremely helpful to the verification engineer, who will however need to give careful consideration to operating experience, as well as to plant modifications and changes in the intent that have occurred since the time of the original reviews.

d)   Operating instructions. A full understanding of a process may only be possible when the operating intent and practices are known. Matters such as start-up and shut down conditions, remote start-up, parallel operation or changes in feed material are covered in these documents and may not be apparent from other information.

In the case of a reciprocating compressor, the operating instructions showed that while the machine was normally protected from overload by a low suction-pressure alarm, it was necessary to run with this alarm disabled during a regeneration cycle. Once recognised, the danger of not having the alarm enabled during normal operation became apparent. Changes to the instrumentation and operating practice were made to ensure that the correct protection was in place for all modes of operation.

3.3   Machine Maintenance and Operation

a)   Operating Experience

Information on the history of a machine must form part of any assessment. With new installations, information must be available from the manufacturer identifying the possibilities of serious failure and the measures that have been adopted to avoid them.

In the case of existing equipment, where the machines are complex or where there is limited experience of them within the organisation, information should be sought from the manufacturer on the possibilities of serious failure, particularly events which could threaten the integrity of the equipment or lead to unacceptable situations.

The operating history of the equipment should highlight the occurrence of serious failures and all classified incidents must be reported.

The information gained here will give an indication as to how well suited the machine is to its duty and – where satisfactory experience exists and is documented – may obviate the need on existing equipment for checks on piping loads and inspections.

b)   Operating Instructions

The scope of the operating instructions covers items essential to the protection and the reliable operation of the machine. However, from observation, experienced operators can usually develop ways of carrying out simple checks and inspections of a machine which are more effective than those given in operating instructions. This experience is invaluable; it should be recorded and steps should be taken to ensure that it is not lost in the event of any restructuring of the operating teams. In many ways these checks can prove superior to the use of sophisticated instrumentation.

Within the organisation, these measures may be strengthened by the integration of the operating and maintenance teams. This gives an opportunity for improved surveillance gained from a better level of technical knowledge of the machine.

c)   Maintenance Instructions

Planned maintenance instructions must be based on a considered maintenance policy in which all required checks are fully identified. Reliability analysis of the machine and its systems will identify the critical parts of the machine and will aid the determination of the frequency of maintenance needed if the required level of reliability is to be sustained. Given that the required level of safety is attained, maintenance policies must also be selectively applied for optimum cost effect.

For some items reliability analysis will show that the application of a given maintenance action should be usage-based (e.g. after processing a given volume of material) or time-based (which is usually appropriate for safety devices). In other cases, the need, or otherwise, for maintenance can be predicted by regularly monitoring appropriate parameters of operation (e.g. achieved pressure) or condition (e.g. bearing vibration amplitude), thus avoiding premature shutdown for inspection or overhaul. Where failure is not critical (where the machine has a standby, for example) maintenance can be on a breakdown basis.

Whatever the maintenance policy, the instructions must be clear about the inspections required at the time of any overhaul.

In all cases, good maintenance practice, carried out by appropriately trained personnel, is assumed. Specific checks are identified for critical maintenance activities, where problems have been found to be generic to certain types of machine. Examples of this are the checks on the throttle bush clearance for a hydrocarbon duty pump, checks on foundation bolting, and the assembly of the piston rod to cross-head connection on a reciprocating compressor.

Manufacturer's instructions will be taken account of during the verification and, where the operation described is key to the machine integrity, rigid adherence to particular safety elements will be specified in the verification report. There is scope for a degree of interpretation of such instructions – within the bounds of good maintenance practice – but all repairs and modifications to a machine will need to be authorised.

3.4 Hazard Review

This is a review with a particular focus on aspects of the machine design; in some cases this will require multiple reviews, as there are several specifics to consider. The reviews are generally based on the engineering guides or specifications produced by the operating company. Although in general these can be simple 'yes or no' reviews, judgement is required when considering new-technology equipment or assessing mitigating circumstances for equipment installed prior to the review.

6.3.5   *Containment Review*

To facilitate this review a series of structured checks on the machine design should be devised. These should take account of the majority of containment situations found on common machines so as to be adequate for most of them. However, these reviews may not be enough; experience shows that the verification engineer must then make a judgement on the overall design based on experience and analysis of the machine. This examination should be done at the end, because the items of information will only come together when the full design of the machine is understood, its history is known, and the protective system is defined.

A modification of an agitator was under review. The modification only involved changes around the bearing support and, as far as it went, was satisfactory. However, the verification engineer continued to have concerns about the equipment, because the restraint of the agitator depended on only three bolts; operating history showed that there was evidence of the shaft moving in the coupling – designed to be a sliding fit – and failure of such restraining bolts had been found on other installations.

The consequences of failure would have been severe; there was a possibility that the shaft might then drop through the seal, resulting in an unconstrained release of flammable vapour from a major vessel. Redesign of the modification provided both a way of addressing the original intent – which was to improve bearing life – and a means of eliminating the threat to containment.

7 OPERATIONS AND MAINTENANCE

INTRODUCTION

Hazards to a designer are perceived but to a plant operator they become realities.

The plant operator must first understand what was in the mind of the designer. He must then relate it to the reality of the plant and the resources available to him. The operating situation may be different to that envisaged by the designer. The resources in maintenance, both men and materials and the actual skills of the operators will need to be assessed against the tasks needed to be carried out.

The risks to safety have to be managed throughout the life of the plant.

The probability of failure increases with time due to complacency of operators and plant deterioration. The measures needed to ensure safety must be audited on a regular basis so that methods and procedures are kept up to date. Improvements to plant and machinery must be carried out with proper integration into the existing safety systems.

1 Handover

Before completion, and prior to operation, the legal documents for the plant must be formally handed over to the client so that staff training and maintenance planning can based on their contents. Meetings must be held with all involved parties so that inherent risks and safety provisions are fully understood. The legal documents as prescribed are:

a)   The Safety Plan

This is the document listing all the actions which have been taken during the design and construction phase of the project in accordance with the CDM Regulations. The plan will need to be extended to include all the actions needed to ensure safe operation of the plant so as to safeguard the health and safety of the workers and the general public.

b)   The Safety File

This contains a compilation of the design features provided for the health and safety of the plant workers resulting from the execution of the safety plan during design and construction, as required by the CDM Regulations. This is in effect all of the work as outlined in the Hazard Assessment (Section 4) of the guide.

c)   The Technical File

The Technical File, in accordance with the Machinery Directive (see Section 1, *Legal Requirements*) is required to be held only by the vendor. The regulations do not require it to be issued to the client.

For process plant machinery, however, it is imperative that the client is made aware of the contents of the technical file, and it should be specified for supply as part of the purchase order.

Process machinery supplied in accordance with API specifications usually has to include Operating and Maintenance Manuals and QA documentation. These normally contain:

- Operating instructions;
- Verification reports;
- Maintenance instructions and inspections;
- Manufacturing records and data sheets;
- Design limits of the machine;

56

- Identification of hazards and failure modes;
- Safeguards provided to ameliorate the effects of hazards and failures on health and safety;
- Schedule of protective systems;
- Availability and reliability data.

The required content of the Technical File is similar to the above.

For the process industries, it would seem that the technical file and the technical manuals should be made one and the same, to be held by the client as well as by the vendor.

.2 Reliability-Centred Maintenance (RCM)

Any machine, however well designed, must be maintained to ensure reliable safe operation. CDM regulations require maintenance to be considered from the earliest stages of a project. Requirements and targets must be established at the start of a project when budgets and resources are being allocated. This ensures that maintenance facilities will be provided for in the plant design.

RCM is an ideal tool which, if applied at the design stage, will lead to optimising planned maintenance and managing the risk to safety in operation.

In the reliability assessment of the machine, the various failure modes will have been identified. When the failure modes have been identified, the steps needed to control any hazards can then be proscribed. Very often this will lead to the addition of critical safety devices. Some of these devices will be passive in nature – not needed for normal operation, but needed under emergency conditions. Maintenance operations are usually production-orientated, and devices that do not affect production can easily be overlooked. Unless singled out as safety critical, these devices will not command the attention warranted for their maintenance.

RCM ensures the identification of all safety-related devices and enables maintenance planning for safe and reliable operation to be based on the optimum mix of:

- scheduled maintenance;
- predictive maintenance;
- breakdown maintenance.

RCM analysis consists of:

1. defining the functional structure of the machine and ranking its subsystems in the order of their effect on safety;
2. identifying items where failures have a significant effect on safety and reliability;
3. identifying for each item the significant failure modes, their likely causes and possible means of detection;
4. for each failure mode, selecting the optimum maintenance procedure for minimising the risk to safety and reliability;
5. setting up the maintenance plan;
6. implementing the plan and establishing an effective feedback loop to allow periodic adjustment and update.

7.3   Operations and Maintenance

It is the responsibility of the client (or other such designated competent person as appointed by the client) to ensure that the machine system is operated and maintained so as to stay within its design limits and to ensure that the protective systems are in place and operational. This requirement includes the maintenance of integrity through the application of appropriate operating procedures, maintenance policy and practices. Reports and records must also be maintained in accordance with recognised QA/QC procedures for verification.

The Client must ensure that personnel who operate, maintain, test or inspect critical machine systems have received adequate training and instructions to enable them to carry out their duties safely and effectively, both for normal operation and in the event of an emergency. Operator training is critical as loss of control in the first five minutes of an incident can rapidly escalate to disaster. (See Figure 6.)

In particular, process operators should be made aware of the importance of protective systems, their safety implications and the dangers which can result from their misuse. Maintenance personnel should be aware of the importance of standards of workmanship and attention to detail in maintaining critical machine systems. Handover from a live operating plant to a maintenance activity entails a significant level of risk. A rigorous permit-to-work system, with appropriate enforcement and monitoring, is clearly necessary. To ensure this requires proper training and competent management.

The client should ensure that reports of failures of critical machine systems are sent to an appropriate Designated Authority who must maintain a data base and inform other users. The Designated Authority can be set up in the client's own organisation and/or with the vendor as appropriate. Maintenance procedures and schedules must be routinely reviewed in the light of up-to-date information. Reference should also be made to the PUWER requirements.

.4 Inspection

The 'Responsible Person', as identified in the safety file – usually the equipment package vendor, plant designer or operator – will hold recommended frequencies of inspection etc. in the Schedule of Protective Systems and Inspection Frequencies. These schedules and frequencies shall be reviewed after every overhaul or examination of the machine system, and approved for adjustment as necessary. (Refer to PUWER, Regulation No.6).

Reports providing evidence that the specified periodic activity for inspecting, examining or testing equipment or systems has been accomplished shall be maintained in the equipment file and monitored by the In-service Inspection Section at appropriate intervals.

7.5 System Audits

Audits should be carried out to ensure that management systems which have been established to assure safe operation are adequate and effectively maintained. The frequency of audit depends on the process, on the machine hazards and also on the frequency of turnover of experienced staff. In the process industries a frequency of once every three to five years is normally accepted as optimum for detailed system audits, although local audits by plant staff should be done more often to ensure that the actual practice follows the intended one.

Audit protocols will normally include the following checks:

- On responsibilities of key staff: ensuring that key technical staff understand their responsibilities and are trained and competent in their role.
- On written procedures: agreed practice needs to be recorded in local plant procedures so that all staff are clear about the management systems which have been established to ensure safe operation and the control of hazards. These procedures will include any statutory requirements.
- On examination and notification: systems established to inspect and test equipment and any notification system to identify that the examination is due.
- On documentation and records: files relating to the duty, design limits, maintenance and operation requirements and examination records.

.6 Management of Change

.6.1 Modifications

The Machinery Directive requires that any modification requires the same procedure as for a new machine. Both API and the HSE have identified the need to ensure that changes made to chemical and petrochemical plants, either in design, construction or use, shall undergo a reasonable proving or validation process.

Modifications as applied to a chemical or petrochemical works are usually for changing, expanding, or altering feedstocks, etc. and, as such, may easily result in breaching the original design parameters and concepts. Because of this, API and OSHA – as well as the HSE – are very concerned and aware of the possible hazards generated by such changes and have devised certain codes addressing 'the management of change'.

The aim of this section is to aid compliance with the intent of the various legislative requirements, with particular focus on the duties of the contractor or user.

.6.2 The Assessment of Change

The assessment of change should cover all modifications (as noted in Section 2), changes of materials, changes of specification, temporary installation, bypasses, etc. which may affect the integrity of the plant or protection systems or violate in some way the mechanical or other adequacy of equipment for its specified duty. It shall also consider any changes to instrumentation, electrical or software control systems which may affect the integrity of the process or utility operations.

.6.3 Compliance with Requirements

All changes which will involve machinery must be subject to the requirements as given in this guide. Any existing CE

61

mark will thereby become invalidated, the technical file will need to be revised and a new CE mark applied. The technical file is to be updated by the user, with contact and feedback to the original manufacturer(s).

In the design and construction of the modification, a hazard assessment with a safety plan and a safety file will be required, as explained in this guide.

7.7 Approval of Modifications

7.7.1 Checks and Reviews

A Safety, Health and Environmental (SHE) assessment and an Essential Health and Safety Requirement (EHSR) assessment must be carried out. Whenever possible, these checks must be completed before any modification is carried out. If for some reason the checks are not possible, then as a last default they must be completed within 72 hours of carrying out the modification. This assessment will involve deciding whether a HAZOP study or SHE review of the modification will be necessary.

A HAZOP study or SHE review may be carried out after the modification has been installed if the modification was required in an emergency to mitigate an unsafe situation.

Modifications where a HAZOP study or SHE review is not necessary will depend on the opinion of the reviewers. They must decide whether or not the modification interferes in any way with the integrity of the plant, process or system.

7.7.2 The SHE Assessment Form

The SHE assessment form must be annotated to indicate whether a HAZOP study or an SHE review is necessary. This form shall also record:

- engineering modifications;
- control or software modifications;

- chemical or composition changes;
- existing safety and technical files, if available.

The form shall also reflect or give attention to:

a) the plant or area of change;
b) the section of the plant or area, item or tag number of item;
c) the date or proposed date of the change;
d) details of the change;
e) the reason for the change, and its originator;
f) whether SHE or HAZOP is required;
g) problems caused by the change (or new factors that it has created) and ways of minimising their impact.

.7.3   *Initialling and Dating*

The SHE assessment form must be initialled and dated by the following:

1. The area facilities engineer
2. The process engineer
3. The loss prevention engineer
4. The control and instrumentation engineer, where applicable.

.7.4   *Recording of Details*

When changes are enacted, the following details shall be recorded as soon as possible:

a) HAZOP study or SHE review comments and recommendations;
b) Details of the changes – including PFDs, P&IDs, piping drawings, control and instrumentation details, materials data;
c) Safety and technical files and all documentation as required by this guide.

7.8 **Design Standards**

All modifications and changes should be designed, installed and tested to recognised codes and statutory standards where appropriate.

8 REFERENCES

FURTHER READING

1   Carter, A.D.S., *Mechanical Reliability,* 2nd edition, Macmillan, London 1986, ISBN 0 333 40587 0.

2   Bloch, H.P., and Geitner, F.K., *Machinery Reliability Assessment*, Van Nostrand Reinhold, New York 1990, ISBN 0 442 23279 9.

3   Andrews, J.D. and Moss, T.R., *Reliability and Risk Assessment*, Longman Scientific, Harlow 1993, ISBN 0 582 09615 4.

4   Davidson, J. and Hunsley, C. (editors), *The Reliability of Mechanical Systems, 2nd Ed*, Mechanical Engineering Publications, IMechE, London 1994, ISBN 0 85298 881 8.

5   Thompson, G., *The Comparative Design Evaluation of Pipe Joints with Respect to Reliability*, IMechE. Seminar on Valve and Pipeline Reliability, Manchester, 1993.

6   O'Connor, P.D.T., *Practical Reliability Engineering (3rd Ed, Revised)*, Wiley, Chichester 1995, ISBN 0 471 96025 X.

7   Moubray, J., *Reliability-centred Maintenance*, Butterworth-Heinemann, Oxford 1991, ISBN 0 7506 0230 9.

8   Lees, F.P., *Loss Prevention in the Process Industries*, Butterworth-Heinemann, Oxford, ISBN 0 7506 1547 8.

9   Kletz, T., *Learning from Accidents*. Butterworth-Heinemann, 1994, ISBN 0 7506 1952.

EUROPEAN DIRECTIVES AND EQUIVALENT UK REGULATIONS

The Machinery Directive	Council Directive	89/392/EEC
	Council Directive	91/368/EEC
	Council Directive	93/44/EEC
The Supply of Machinery (Safety Regulations)	1992 UK Reg.	SI 3073
Product Standards – Machinery – DTI Business in Europe		hotline 0117 944 4888
The Use of Work Equipment by Workers at Work	Council Directive	89/655/EEC
The Provision and Use of Work Equipment Regulations	1992 UK Reg.	SI 2932
Electromagnetic Compatibility Directive	Council Directive	89/336/EEC
	Council Directive	91/263/EEC
	Council Directive	92/31/EEC
Electromagnetic Compatibility Regulations	1992 UK Reg	SI 2372

The Low Voltage Directive	Council Directive	73/23/EEC
Electrical Equipment (Safety) Regulations	1994 UK Reg	SI 3260
The Gas Appliances Directive	Council Directive	90/396/EEC
Gas Appliances (Safety) Regulations	1992 UK Reg	SI 711
Simple Pressure Vessel Directive	Council Directive	90/488/EEC
Proposed Pressure Equipment Directive	Council Directive	96/.../EEC
Proposed Pressure Equipment Regulations		

The equivalent laws enacting the above should be obtained from the national body of the European country of interest.

VIDEOS

(Produced by the Mechanical Reliability Committee of the IMechE. and obtainable – cost on application – from the Continuing Education Unit, University of Manchester School of Engineering, The University, Oxford Road, Manchester M13 9PL, UK).

1 *Learning from failures*
Weibull reliability analysis, technique and applications.

2 *Exploring failure consequences*
Failure mode and effect analysis, technique and applications.

3 *Dissecting system failures*
Fault tree analysis, technique and computerised application.

PAPER

Moss, T.R., and Andrews, J.D,, *Reliability Assessment of Mechanical Systems*, to be published in Part E, Proceedings of the Institution of Mechanical Engineers.

PUBLIC DATA SOURCES

RELDAT ™ from AEA Consultancy Services, Risley, Warrington, UK.

OREDA (1992) Offshore Reliability Data Handbook. Det Norske Veritas Industri, Norge AS, DNV Technica.

EIReDA, European Industry Data Handbook, Editions SFER, Paris.

ORGANISATIONS

Health and Safety Executive, Baynards House, 1 Chepstow Place, London, W2 4TF, UK.

Commission of the European Communities, DG X 1. E.1.

Safety and Reliability Society, Clayton House, 59 Piccadilly, Manchester, M1 2AQ, UK.

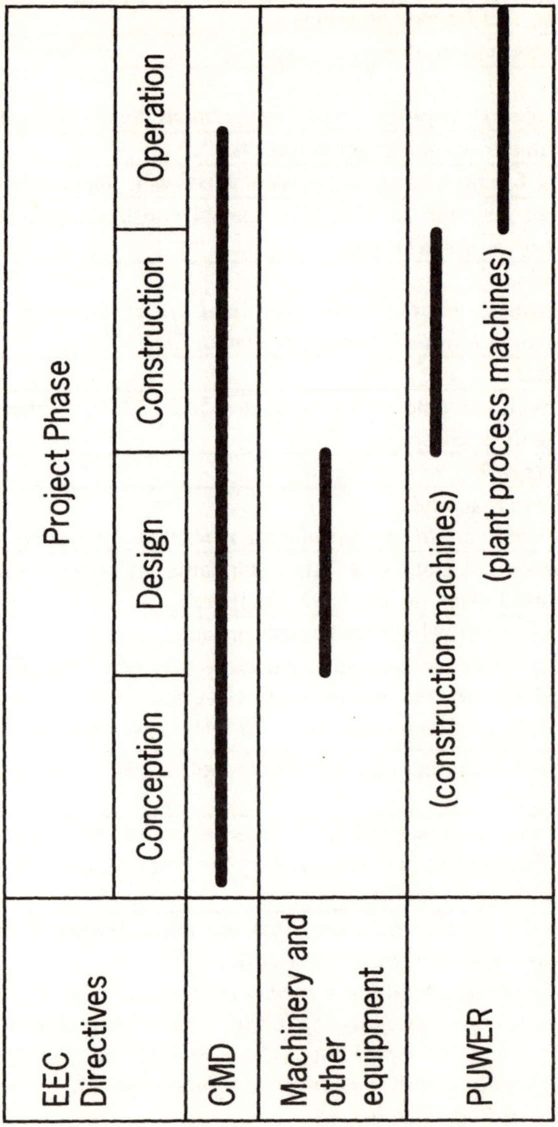

Figure 1 Legal framework

CONTENTS

Name and business address of Responsible Person.
Title of machine or its designation of series or type *(e.g. model number)*.
Description of machine. *Intended use and limits of use e.g. scope etc. Full list of variations (family of machines).*
Statement whether machine is unique or series-produced.
Statement whether machine functions on its own or is combined with other machinery.
Statement whether or not machine is subject to special attestation procedures *(i.e. listed in Schedule 4 of the Regulations).*
Overall drawing of machine. *General assembly drawings, photographs, technical artist impressions to uniquely identify the machine in relation to the technical file.*
Drawings of control circuits.
Full detailed drawings, with any calculation notes, test results, etc. to check the conformity of machine with EHSRs. *Only as so far as it affects safety.*
List of EHSRs that apply to machine. *Complete list of EHSRs which could apply and state where not applicable.*
Lists of transposed harmonised standards, other standards and other technical specifications that were used when machine was designed.
Description of methods adopted to eliminate hazards presented by machine. *It is important that this is a description and not a mere statement. Ideally there should be a justification of why the methods are adequate. A useful way of achieving this is to back this justification up with a Risk Assessment of some form.*
Technical reports or certificates obtained from a competent body or laboratory *(e.g. EMC test report).*
If declaring conformity with a transposed harmonised standard, any supporting technical report giving results of tests carried out by the manufacturer, or at his choice, or by a competent body or laboratory.
For series manufacture, the internal measures that are implemented to ensure that future machines remain in conformity with EHSRs. *Reference to a Quality System is useful but a resume' is needed pointing out the key features of the system, such as Drawing Office procedures, Production Control, Inspection etc. which relate to this. There should also be a clear summary of how modifications will be approved internally and how the interaction with the Notified Body is achieved.*
Copy of instructions for machine. *Including proposed Declaration of Conformity and any specimen copies of test certificates which would be supplied to customers*
Copies of technical literature provided to customers.

Figure 2 - Technical File Contents

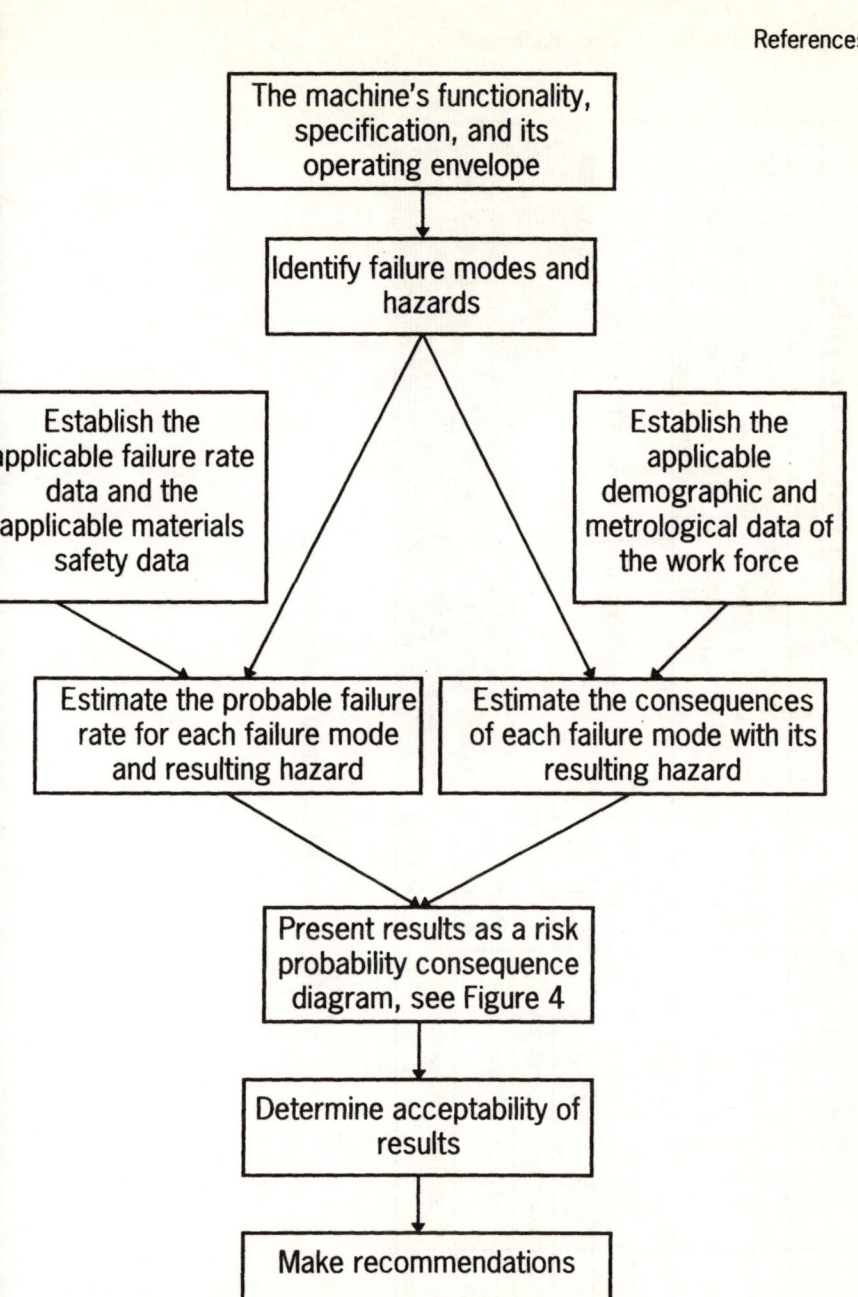

Figure 3 Flow diagram for risk assessment

Severity rating	People Injury	Assets Damage	Environment Effect	Public Relations Impact	A Unknown in the world	B Unknown in the industry	C Has occurred in the industry	D Many times a year in the industry	E Many times a year at the location
0	None	None	None	None					
1	Slight	Slight	Slight	Slight					
2	Minor	Minor	Minor	Limited					
3	Major	Localised	Localised	Considerable					
4	One fatality	Major	Major	Major National					
5	Many fatalities	Extensive	Massive	Major International					

CONSEQUENCE / INCREASING PROBABILITY

Regions: SAFETY MANAGEMENT ACTION; RISK REDUCTION MEASURES REQUIRED; INTOLERABLE

Figure 4 - Risk Classification Matrix

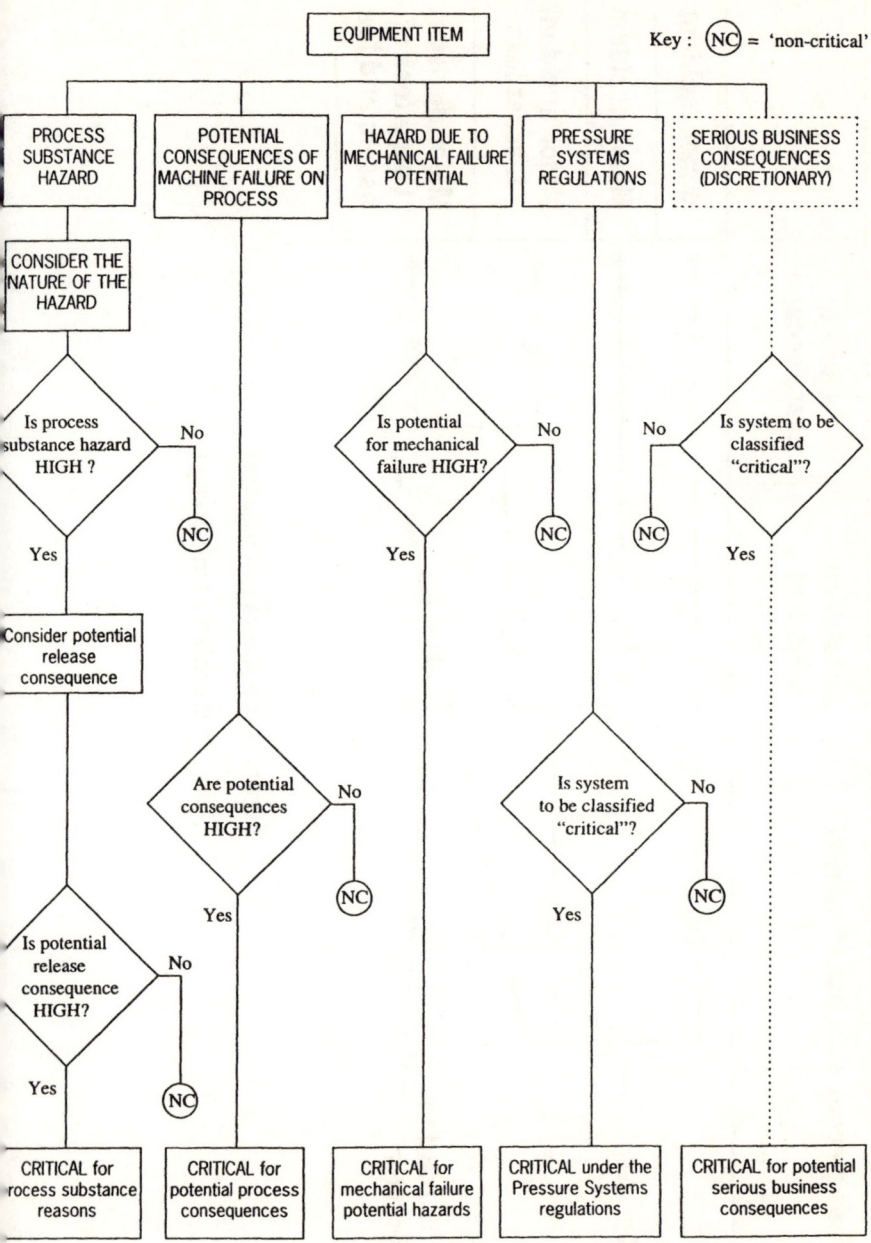

Figure 5   Machines Classification Diagram

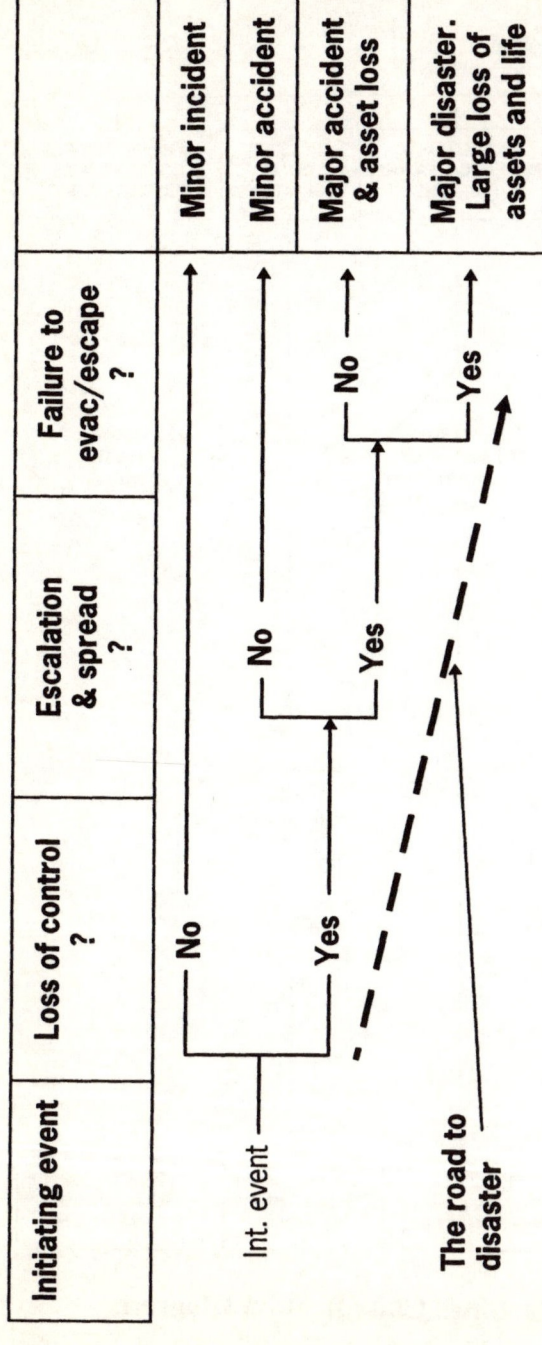

**Figure 6 Escalating to disaster
(J Strutt Cranfield University)**

9

APPENDICES

APPENDIX A: DEFINITIONS

API

American Petroleum Institute.

Availability

The ability of an item (under combined aspects of its reliability, maintainability and maintenance support) to perform its required function at a stated instant of time or over a stated period of time.

CDM

Construction (Design & Management) Regulations.

Conformity Assessment

A review of the Machinery Directive requirements in order to identify the actions needed to conform with the regulations.

Critical Machine

A machine which would cause an unacceptable situation should the machine or its protection system fail.

EEA

European Economic Area.

EEMUA

Engineering Equipment and Materials Users' Association.

Essential Health And Safety Requirements (EHSRs)

This is a list of requirements such as instructions, controls, adequate lighting and so on, applicable to a whole range of possible hazards such as mechanical and electrical problems and noise.

Fog Index

A method for measuring readability based on the length of a sentence and the complexity of the words used.

HAZAN

Hazard Analysis.

HAZOP

Hazards and Operability Study.

HSE

Health and Safety Executive.

IEC

International Electrotechnical Commission.

Incorporated Machine

In Machinery Directive (Reg. 23), an 'incorporated machine' is machinery that:

76

(a)   *is intended for:*
      (I) incorporation into other machinery, or
      (ii) assembly with other machinery, to constitute relevant machinery.
(b)   *cannot function independently: and*
(c)   *is not interchangeable equipment.*

Inspection

The process of measuring, examining, testing, gauging or otherwise comparing the unit with the applicable requirements.

LEL

Lower Explosive Limit

Machine

In Machinery Directive (Reg. 4) a 'machine' is defined as:

(a) an assembly of linked parts or components, at least one of which moves, including – without prejudice to the generality of the foregoing – the appropriate actuators, control and power circuits, joined together for a specific application, in particular for the processing, treatment, moving or packaging of a material;

(b) an assembly of machines, that is to say, an assembly of items of machinery as referred to in paragraph (a) above, which, in order to achieve the same end, are arranged and controlled so that they function as an integral whole, notwithstanding that the items of machinery may themselves be relevant machinery and accordingly severally required to comply with these regulations; or

(c) interchangeable equipment modifying the function of a machine which is supplied for the purpose.

This definition is amended to include: *safety components,* such as electro-sensitive devices designed to detect persons to ensure their safety, logic units which ensure the safety functions of bi-manual controls, roll-over protection structures and falling object protection structures.

Machine Classification

In a given plant or process it is normal practice to subdivide the plant into a number of units or sub-units as convenient. The conformity assessment can be carried out on all of them, so that any machinery identified can be classified in order to prioritise the hazard potential of each unit and to identify its critical components.

Maintenance

The combination of all technical and associated administrative actions intended to retain an item in, or restore it to, a state in which it can perform its required function.

OSHA

Occupational Safety and Health Administration.

PFD

Process Flow Diagram (sometimes called a 'flowsheet').

P&ID

Piping and Instrumentation Diagram (also known as an Engineering Line Diagram).

PUWER

Provision and Use of Work Equipment Regulations 1992.

RAMS

Reliability, Availability, Maintenance and Safety.

Registration

The process of machine classification, identification, verification, recording and inspection to maintain the viability of the process.

Reliability

The probability that a machine will perform its prescribed duty without failure for a given time when operated correctly in a specified environment.

Reliability-Centred Maintenance (RCM)

A maintenance strategy drawn up via a structured framework of analysis aimed at ensuring the attainment of a systems inherent reliability. This results in a criticality ranking of maintenance operations based on component contribution to the overall reliability.

Safety Integration

The elimination of the identified risks by good design instead of by 'add on' guards and safety devices.

SHE

Safety, Health and Environmental review.

Type Test

The verification, by testing and approval by a competent or notifiable authority, that an inactive unit meets all safety requirements.

UEL

Upper Explosive Limit.

Work Equipment

Any machinery, appliance, apparatus or tool and any assembly of components which, in order to achieve a common end, are arranged and controlled to function as a whole.

APPENDIX B: MATERIALS SAFETY DATA SHEET
SOME COMMON HAZARDS

MATERIAL SAFETY DATA SHEET	PREP BY			
	APPD BY			
	DATE			
	ISSUE	1	2	3

SECTION 1– MATERIAL IDENTIFICATION

MANUACTURERS NAME:	
CHEMICAL NAME:	
TRADE NAME:	
CHEMCIAL FAMILY:	MOLECULAR WEIGHT:
FORMULA:	

SECTION 2 – HAZARDOUS INGREDIENTS (SEE NOTE 2)

	%	TLV-TWA	TLV-STEL	TLC-CL

SECTION 3 – PHYSICAL DATA

BOILING POINT:	FREEZING POINT			
VAPOUR PRESSURE @ 20 C:	SPECIFIC GRAVITY (WATER=1)			
VAPOUR DENSITY (AIR=1):	PERCENT VOLATILES BY VOLUME:			
SOLUBILITY IN WATER:	EVAPORATION RATE:			
APPEARANCE AND ODOUR:				
CHEMICAL NAME				

MATERIAL SAFETY DATA SHEET	PREP BY			
	APPD BY			
	DATE			
	ISSUE	1	2	3

SECTION 4– FIRE AND EXPLOSION HAZARD DATA

FLASH POINT:			
FLAMMABILITY LIMITS IN AIR (% BY VOL):	LFL=	UFL=	
AUTOIGNITION TEMPERATURE:			
EXTINGUISHING MEDIA:			
SPECIAL FIRE FIGHTING PROCEDURES:			
UNUSUAL FIRE FIGHTING PROCEDURES (eg DUST):			
FIRE AND EXPLOSION RISK:			
MAXIMUM RATE OF PRESSURE RIST/St CLASS:			

SECTION 5 – HEALTH HAZARD DATA

THRESHOLD LIMIT VALUES:	TLV-TWA=	TLV-STEL	TLV-CL=
SAX HAZARD RATING:			
EFFECTS OF OVEREXPOSURE			
EMERGENCY AND FIRST AID PROCEDURES:			

SECTION 6 – REACTIVITY DATA

STABILITY::	
CONDITIONS TO AVOID:	
INCOMPATABILITY, CORROSIVITY (MATERIALS TO AVOID)	
HAZARDOUS DECOMPOSITION PRODUCTS:	
HAZARDOUS POLYMERISATION:	
CHEMICAL NAME:	

84

MATERIAL SAFETY DATA SHEET	PREP BY			
	APPD BY			
	DATE			
	ISSUE	1	2	3

SECTION 7– SPILL OR LEAK PROCEDURES

STEPS TO BE TAKEN IF MATERIAL RELEASED OR SPILLED:

WASTE DISPOSAL METHOD:

SECTION 8 – SPECIAL PROTECTION INFORMATION

RESPIRATORY PROTECTION, SPECIFY TYPE:
VENTILATION:
LOCAL EXHAUST:
SPECIAL:
MECHANICAL REQUIREMENTS (eg SPARK PROOF):
OTHER:
PROTECTIVE GLOVES (STATE MATERIALS):
EYE PROTECTION:
OTHER PROTECTIVE EQUIPMENT:

SECTION 9 – SPECIAL PRECAUTIONS

PRECAUTIONS TO BE TAKEN IN HANDLING AND STORING

PRECAUTIONARY LABELLING

OTHER PRECAUTIONS

MATERIAL SAFETY DATA SHEET	PREP BY			
	APPD BY			
	DATE			
	ISSUE	1	2	3

CHEMICAL NAME

SECTION10 - MISCELLANEOUS SECTION FOR ADDITIONAL INFORMATION

1. REFERENCES (ENSURE THAT LATEST EDITIONS ARE CONSULTED)

　　(1A)　SAX-DANGEROUS PRERTIES OF INDUSTRIAL MATERIALS - (6TH EDITION)

　　(1B)　FIRE PROTECTION GUIDE ON HAZARDOUS MATERIALS - NFPA-(1985 EDITION)

2. TLV-TWA　　TIME WEIGHT AVERAGE REFERENCED TO A 8 HOURS EXPOSURE

　TLV-STEL　SHORT-TERM EXPOSURE LIMIT REFERENCED TO A 10 MINUTES EXPOSURE

　　TLC-CL　　CEILING CONCENTRATION SHOULD NOT BE EXCEEDED EVEN INSTANTANEOUSLY

SOME COMMON HAZARDOUS MATERIALS

NAME	FORMULA	STATE	HAZARD	CONTROL	NOTES
Air	N2 O2	pressurised gas	death/injury	Safety valve Training/ education	Rupture of container results in explosion, rupture can be from over pressure caused by maloperation, weakening as a result of corrosion or overheating in a fire. Detachment of a loose fitting results in a lethal missile Jet impingement can penetrate skin into blood vessel and cause heart failure due to air bubbles in the blood stream
		pneumatic conveying	fire/ explosion	segregation from ignition sources	when used for the movement of powders, which in themselves are harmless can become hazardous. e.g. :- flour and air can explode.
Oxygen	O2	gas	fire	segregation from ignition sources	Highly reactive - even metal will burn with intense heat only needs a low energy source, such as an impact to ignite. Denser than air.
Nitrogen	N2	gas	will not support life	Training/ education	Commonly used to exclude air
Carbon monoxide	CO	colourless gas without odour	fire/ explosion	segregation from ignition sources	LEL 12.5% UEL 74% Auto ignition temp 608°C Danger of combustion when exposed to a flame.
			Death	ventilation/ gas detection	Asphyxiation . Due to affinity with haemoglobin low concentration causes dizziness , tiredness and headache.
Carbon Dioxide	CO2	gas	Death	Training/ education	Commonly used as a fire fighting medium to suppress combustion Need to evacuate in the event of release. Death caused by lack of oxygen .

SOME COMMON HAZARDOUS MATERIALS(continued)

NAME	FORMULA	STATE	HAZARD	CONTROL	NOTES
Halon (Green equivalents)		gas	Death or brain damage	Training/ education	Commonly used as a fire fighting medium to suppress combustion Has the advantage of needing a lower concentration than CO2. Will therefore support life, however will break down into toxic products due to the heat of combustion from a fire. Breathing of toxic gas causes brain damage.
Hydrogen sulphide	H2S	gas	Toxic inflammable explosive	gas detection	Occurs naturally in natural gas and crude oil and is also used in industrial processes. Heavier than air and so will accumulate in low lying areas. Toxic, causes irritation and respiratory paralysis. In low concentrations smells of bad eggs ,exposure will kill the sense of smell. 10 ppm is dangerous.
Phosgene	COCL2	gas	Highly toxic	ventilation breathing apparatus	Used for a wide range of industrial processes for making dyes , pharmaceuticals , etc. etc. 0.1 ppm is dangerous. heavier than air.
Diesel fuel		liquid	benign	Keep away from fire personal hygiene	Will sustain a fire but will not easily ignite Contact with skin can cause dermatitis.
Sulphuric acid	H2SO4	liquid	highly reactive destroys human tissue	Containment of all spills protective clothing safety eye wash and showers	Widely used for industrial processes. Although non combustible could cause a fire after coming in contact with wood or similar substances. When diluted with water and after reacting with metals will produce H2 , which can explode. Fumes are dangerous to be inhaled and the liquid is dangerous to flesh and tissue.

88

AMELIORATION OF HAZARDS

HAZARD	ACTION	NOTES
Fire	segregation	separate sources of ignition from inflammable materials, detect inflammable gas before inflammable /explosive concentrations are reached and isolate the source of gas leak.
	detection	sense light emission of flames, heat radiation, rate of temperature rise.
	escape	provide clearly marked , unblockable, multiple means of escape.
	control	provide means of extinguishing and control from spreading the fire.
	contain	contain spread of inflammable materials and provide means of isolation such as fire proof doors.
	rescue and succour	trained teams , protective clothing , breathing apparatus , means of communication, medical facilities.
enclosed spaces ,pits and cellars	training	there is always danger of asphyxiation or fainting, fire or explosion from accumulated gas. Such spaces must be avoided in design or provided with adequate ventilation
	access	where entry is required , adequate facilities, such as opening size, platform, stairs etc., should be provided to allow entry with all safety equipement on, and to allow rescue if required. Uncontrolled access should be prevented by locks and permit to work procedures.
	testing	Atmospheres must always be tested before entry
	rescue	Standby help with breathing apparatus must always be at hand before entry. The person entering should have safety ropes for retrieval.
Moving parts	separation	Safety guards and protective clothing with no items that could penetrate the guards such as long hair or loose clothing.
	secure	safety devices and procedures to prevent motion before removal of guards and entry
Emergency shut down	containment, isolation and release	contain, isolate and or dispose of all sources of dangerous fluids . Vent all pressurised containers. controlled stop of all operations
Electricity/radiation	separation	mechanical separation and shielding.
toxic gas and other hazards	isolation	Work permit systems, i.e., procedures for making safe and ensuring isolation of the danger before allowing work to proceed.
	detection	Monitoring of a safe environment both before and during work operations.

APPENDIX C: HEALTH AND SAFETY COMPLIANCE CHECK LIST

The Supply of Machinery
(Safety) Regulations 1992 SI 3073
Schedule 3
Essential Health and Safety Requirements Relating
to the Design and Construction of Machinery

Compliance Checklist

PROCEDURE : REVIEW THE EQUIPMENT AGAINST THE QUESTIONS IN THE CHECKLIST.

IF ANY SECTION IS NOT APPLICABLE WRITE N/A IN THE APP COLUMN.

IF THE ANSWER TO ANY QUESTION IS "NO" THEN THE DESIGN SHOULD BE ALTERED TO COMPLY WITH THE REQUIREMENT.

TOPIC	APP	FILE REF
1.1 GENERAL REMARKS		
1.1.2 PRINCIPLES OF SAFETY INTEGRATION		

a) Is the machine fit for function.

b) Can it be operated and maintained without putting people at risk.

c) Have all the uses to which the machine could reasonably be expected to be put been considered.

93

1.1.3 MATERIALS AND PRODUCTS		

a) Are all materials used in the equipment and produced by the equipment, free from health risks.

b) Have adequate provisions been made for filling/draining fluids from the equipment.

1.1.4 LIGHTING		

a) Is any integral lighting required on the equipment and if so has it been provided.

1.1.5 DESIGN OF MACHINERY TO FACILITATE ITS HANDLING		

a) Has the equipment sufficient lifting points or component geometry to achieve a safe and damage free lift of

i. the complete unit.

ii. component parts over 10kg which will be lifted during assembly/disassembly.

1.2 CONTROLS		
1.2.1 SAFETY AND RELIABILITY OF CONTROL SYSTEMS		

a) Is the control system suitably protected to withstand the expected environment and use.

b) Is the controls system designed such that errors in control logic do not lead to dangerous situations.

1.2.2 CONTROL DEVICES		

a) Are control devices

   i. clearly visible and identifiable.

   ii. positioned for safe operation and without time delay or ambiguity.

   iii. designed so that movement of the control is consistent with its effect.

   iv. located outside the danger zone unless this is a necessary requirement.

   v. fitted with all indicators required for safe operation which are visible from the control position.

b) Can the danger zones be viewed and confirmed free of persons from the main control position. If not what warning signals are provided to warn persons in the danger zone of machinery starting.

1.2.3 STARTING		

a) Can the equipment only be started by actuation of a control provided for the purpose (if no then go to question b).

b) If the equipment has several starting controls has a selector been fitted to ensure that these starting controls can be de-selected as required.

1.2.4 STOPPING DEVICE		

a) Normal stopping. Has the equipment been fitted with a stop control

   i. which has priority over any start controls and depowers once the machinery has stopped.

b) Emergency stopping. Has the equipment been fitted with an emergency stop device which

   i. is clearly identifiable and accessible.

   ii. stops the dangerous process as quickly as possible.

   iii. remains engaged upon actuation.

   iv. allows restarting of the machinery (not start the machine) only upon disengagement.

   v. stops all other equipment in an installation which could pose a danger.

   vi. complies with EN 418.

1.2.5 MODE SELECTION		

a) If the equipment has several operating modes can the mode selector be locked in position.

b) Does the mode selector override all other control systems with the exception of the emergency stop.

1.2.6 FAILURE OF POWER SUPPLY		

a) Will the control system prevent the machine from restarting upon interruption of the power supply.

b) Will the control system prevent any other dangerous situations occurring after interruption of the power supply.

1.2.7 FAILURE OF THE CONTROL CIRCUIT		

a) Will the control system prevent the machine from restarting upon failure or damage to a control circuit or logic.

b) Will the control system prevent any other dangerous situations occurring upon failure or damage to a control circuit or logic.

1.2.8 SOFTWARE		

a) Is any software used in the control system user friendly.

1.3 PROTECTION AGAINST MECHANICAL HAZARDS		
1.3.1 STABILITY		

a) Is the equipment stable under all modes of operation either through geometry or additional anchorage points.

1.3.2 RISK OF BREAK-UP DURING OPERATION		

a) Have all components of the equipment liable to break-up been checked in line with WPL design procedures.

1.3.3 RISKS DUE TO FALLING OR EJECTED OBJECTS		

a) Have any precautions to be taken to prevent risks from falling or ejected objects.

1.3.4 RISKS DUE TO SURFACES, EDGES OR ANGLES		

a) Is the equipment free from rough or sharp edges which could cause injury.

1.3.5 RISKS RELATED TO COMBINED MACHINERY	N/A	
1.3.6 RISKS RELATED TO VARIATIONS IN ROTATIONAL SPEED OF TOOLS	N/A	

1.3.7 PREVENTION OF RISKS RELATED TO MOVING PARTS		

a) Has the equipment been constructed such that risks from moving parts are prevented by design or by guards as specified in section 1.4.

1.4 REQUIRED CHARACTERISTICS OF GUARDS AND PROTECTION DEVICES		
1.4.1 GENERAL REQUIREMENT		

a) Are all guards fitted to the equipment

i) of robust construction (refer to EN 953 for guidance).

ii) not easy to bypass.

iii) an adequate distance from the danger zone (refer to EN 294 and SUPDES1 for guidance).

1.4.2 SPECIAL REQUIREMENTS FOR GUARDS		
1.4.2.1 FIXED GUARDS		

a) Are all fixed guards held in place with fasteners which can only be removed by tools.

1.4.2.2 MOVABLE GUARDS	N/A	
1.4.2.3 ADJUSTABLE GUARDS RESTRICTING ACCESS	N/A	
1.4.3 SPECIAL REQUIREMENTS FOR PROTECTION DEVICES	N/A	

1.5 PROTECTION AGAINST OTHER HAZARDS		
1.5.1 ELECTRICAL SUPPLY		

a) Does all equipment comply with the Low voltage directive (SI 3260 1994) where applicable.

1.5.2 STATIC ELECTRICITY		

a) Does the equipment have suitable earthing points or other devices to prevent a build-up of static.

1.5.3 ENERGY SUPPLY OTHER THAN ELECTRICITY		

a) Has any pneumatic, hydraulic, or thermal equipment been suitably designed to eliminate all potential hazards.

1.5.4 ERRORS OF FITTING		

a) Where parts being fitted wrongly could cause a risk, has the design prevented the possibility of this occurring.

1.5.5 EXTREME TEMPERATURES		

a) Have steps been taken to guard surfaces which could be outwith the temperature range of -10°C to 75°C. If not have warning notices been provided (refer to EN 563 for guidance).

1.5.6 FIRE		

a) Has any equipment which is a potential fire risk been designed to eliminate or minimize the risk.

1.5.7 EXPLOSION		

a) Has any equipment which is a potential explosion risk been designed to eliminate or minimise the risk.

1.5.8 NOISE		

a) Has equipment been designed to reduce noise where required for safety or customer specification (see also 1.7.4).

1.5.9 VIBRATION		

a) Has the balance grade specified removed any potential risks from vibration.

1.5.10 RADIATION		

a) Has the equipment been designed to prevent any emission of radiation.

1.5.11 EXTERNAL RADIATION		

a) Has the equipment been designed to be unaffected by any specified radiation source.

1.5.12 LASER EQUIPMENT	N/A	

1.5.13 EMISSION OF DUST, GASES, ETC.		

a) Has the equipment been designed to prevent risk from dust, liquid, vapour, or gas which it produces or handles by either:

i. reducing the amount of substance emitted to safe levels.

ii. providing suitable containment, evacuation, or drainage.

1.6 MAINTENANCE		
1.6.1 MACHINERY MAINTENANCE		

a) Are all adjustment, lubrication, and maintenance points outside danger zones.

b) If any of the above are not outside danger zones has the design reduced any residual risk to a minimum.

1.6.2 ACCESS TO OPERATING POSITION AND SERVICING POINTS		

a) Has the equipment been provided with suitable means of access to all maintenance points, including ladders, catwalks etc. if required.

1.6.3 ISOLATION OF ENERGY SOURCES		

a) Has the equipment a means of isolating it from all energy sources.

b) Are these isolators clearly identified.

c) Can these isolators be locked off to prevent accidental operation.

1.6.4 OPERATOR INTERVENTION		

a) Has the equipment been designed to minimise the need for operator intervention.

1.6.5 CLEANING OF INTERNAL PARTS		

a) Can all internal parts which could contain dangerous substances be cleaned without risk.

102

1.7 INDICATORS		
1.7.0 INFORMATION DEVICES		

a) Is all information needed to control the equipment clearly identified on the equipment and easily understood.

1.7.1 WARNING DEVICES		

a) If warning devices are provided are they clearly identified and easily understood.

b) Is there a facility for testing the operation of any warning devices.

1.7.2 WARNING OF RESIDUAL RISKS		

a) Are there suitable warning notices on the equipment to warn the operator of any remaining risks to which he could be exposed.

b) Are these warning notices either pictograms or in a suitable language for the country in which the equipment will be used.

1.7.3 MARKING	FOR INFO	

The equipment must be marked with a suitable CE nameplate (refer to SUPDES2 for an example).

1.7.4 INSTRUCTIONS		

a) Does the equipment instruction manual contain the following:

i.   a repeat of the information with which the equipment is marked (usually in the form of a CE pump data sheet, refer to SUPDES3 for an example).

ii. the foreseen use of the equipment as well as how the machinery should not be used.

iii. instructions for putting into service, use, handling (including the masses of major parts), assembly/dismantling, adjustment, and maintenance.

iv. drawings and diagrams necessary for iii). Refer to STD/SDS/0035 for more information.

v.   the equipment sound pressure level and where the sound pressure level is above 85dB(A) the equipment sound power level.

vi. any instructions required for equipment used in an explosive atmosphere.

vii. a suitable translation of at least the safety information and the CE data sheet into one of the languages of the country that the equipment will be put into service.

UPDES 1

XTRACT FROM

N 294 SAFETY DISTANCES TO PREVENT DANGER ZONES EING REACHED BY THE UPPER LIMBS

ROCEDURE FOR DETERMINING THE SAFE DISTANCE

MEASURE OR CALCULATE THE MAXIMUM OPENING IN THE GUARD TO BE EXAMINED.

DETERMINE THE GEOMETRY OF THE OPENING E.G. SLOT.

REFER TO TABLE 5 AND READ OFF THE SAFETY DISTANCE.

THE SAFETY DISTANCE IS THE MINIMUM ALLOWABLE DISTANCE FROM THE GUARD SURFACE TO THE DANGER ZONE.

NOTE THAT TABLE 5 IS ONLY FOR USE WITH PEOPLE ABOVE THE AGE OF 14. IF THE GUARD WILL BE USED WITH PEOPLE BELOW THAT AGE REFER TO EN 294 TABLE 4.

APPENDIX D: DOCUMENTATION RELATING TO A FIRE WATER PUMP

- **CE TYPE NAME PLATE**
- **CE TYPE DATA SHEET**
- **TECHNICAL FILE INDEX**
- **COMPLIANCE CHECK LIST**
- **HAZID and HAZOP DOCUMENTATION (SAMPLE)**

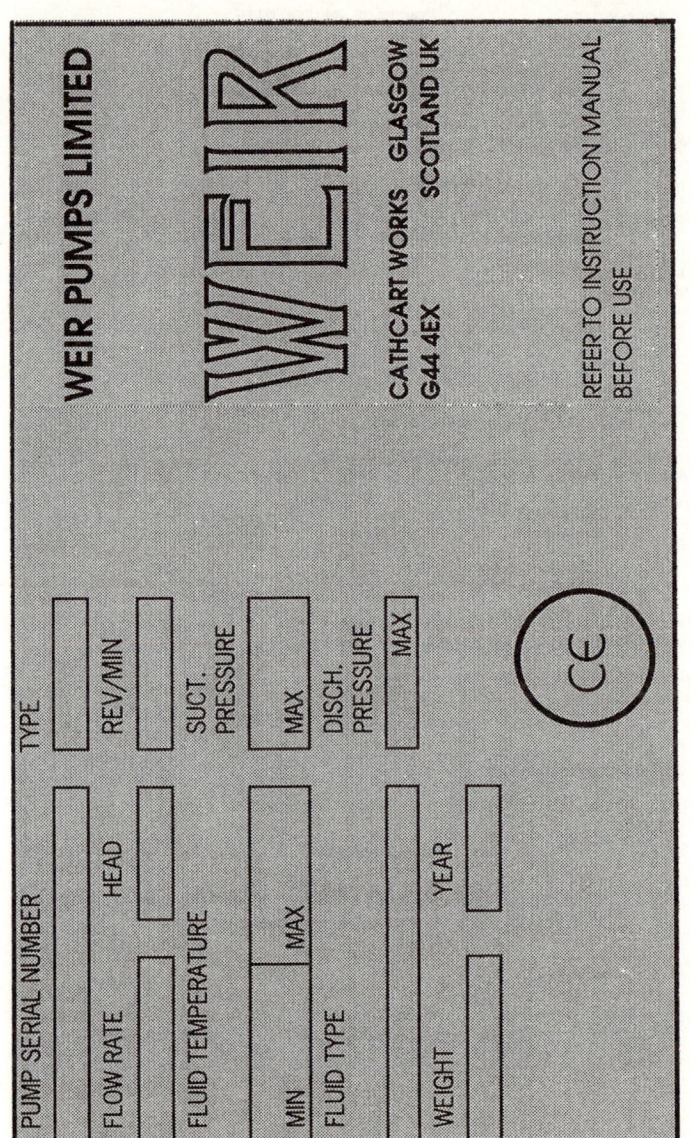

Example of CE type name plate

WEIR PUMPS LTD   PUMP DATA SHEET

ISSUE NUMBER :   02
DATE OF ISSUE :   28-Apr-95
PREPARED BY :   S.BRADSHAW   CONTRACT SPECIFIC DATA

PUMP FRAME SIZE :   HL 50-32-200
PUMP SERIAL NUMBER :   68038-019/20
DRIVER TYPE AND DESIGNATION :   MOTOR D132S

GENERAL ARRANGEMENT DRAWING NUMBER :   A1-604230
SECTIONAL ARRANGEMENT DRAWING NUMBER :   N/A

CONTRACT DUTY HEAD :   36 m
CONTRACT DUTY FLOW :   18 m³/hr
CONTRACT DUTY SPEED :   2890 rev/min
CONTRACT IMPELLER DIAMETER :   181 mm

HYDRAULIC/MECHANICAL LIMITATIONS DATA
ALLOWABLE LIMITS FOR SAFE OPERATION

MAXIMUM PUMP OPERATING SPEED   3000 rev/min

MINIMUM NPSH REQUIRED :   2.4 m
MAXIMUM SUCTION PRESSURE :   1.0 Barg
MAXIMUM DISCHARGE PRESSURE :   4.3 Barg

MAXIMUM CONTINUOUS PUMPED FLUID FLOWRATE :   21.6 m³/hr
MINIMUM CONTINUOUS PUMPED FLUID FLOWRATE :   4.5 m³/hr

MAXIMUM PUMPED FLUID VISCOSITY :   1.519 mm²/s
MINIMUM PUMPED FLUID VISCOSITY :   0.477 mm²/s

MAXIMUM PUMPED FLUID DENSITY :   1000 kg/m³
MINIMUM PUMPED FLUID DENSITY :   983 kg/m³

MAXIMUM OPERATING TEMPERATURE (PUMPED FLUID) :   60 °C
MINIMUM OPERATING TEMPERATURE (PUMPED FLUID) :   5 °C

MAXIMUM OPERATING TEMPERATURE (AMBIENT AIR) :   25 °C
MINIMUM OPERATING TEMPERATURE (AMBIENT AIR) :   0 °C

ALLOWANCES AND LIMITATIONS ON PUMPED FLUID TYPES

FRESH WATER WITH A PARTICLE COUNT OF <150 PPM AND PARTICLE SIZE OF <75 MICRONS
NOT TO BE USED TO PUMP TOXIC FLUIDS
-
-
-
-

ENVIRONMENTAL LIMITATIONS

NOT TO BE USED IN AN EXPLOSIVE ENVIRONMENT
-
-
-
-

DERIVED FROM TEST RESULTS ON SIMILAR EQUIPMENT

SOUND EMISSION DATA

SOUND PRESSURE LEVEL :   76 dB(A)
SOUND POWER LEVEL :   92 dB(A)

The noise is given as a maximum sound pressure level round the equipment, Lp, re 2 x 10^-5 N/m2 at one meter from the
equipment running on its own in the free- field over a reflecting half plane and as a sound power level of the equipment
, Lw, re 10^-12 Watts

Example of a CE pump data sheet.

PRODUCT TECHNICAL FILE

Product designation: SBWM 690 fire pumpset
Contract number: 13168
Issue number: 1

CONTENTS

1)   Safety declaration certificate

2)   The supply of machinery (safety) regulations
     1992 schedule 3 compliance checklist

**THE SUPPLY OF MACHINERY
(SAFETY) REGULATIONS 1992
SCHEDULE 3
ESSENTIAL HEALTH AND SAFETY REQUIREMENT RELATING TO THE DESIGN
AND CONSTRUCTION OF MACHINERY**

COMPLIANCE CHECKLIST

TOPIC	APP	COMMENTS	FILE REF
1.1 GENERAL REMARKS			
1.1.2 PRINCIPLES OF SAFETY INTEGRATION	YES	Complies HAZIDS and HAZOPS used to identify all possible hazards and safety risks.	Appendices 1-8
1.1.3 MATERIALS AND PRODUCTS	YES	Materials used are checked for toxicity and safety in use. Provision for draining is provided where required at engine and gearbox.	Appendix 4 Pump data sheet Appendix 6 Sectional arrangement drawing Parts list GA Drg
1.1.4 LIGHTING	NO	Not required.	
1.1.5 DESIGN OF MACHINERY TO FACILITATE ITS HANDLING	YES	* GA Drg. to indicate unit lifting points. Instruction manual to detail lifting procedure. Erection drawing details how to lift pump.	Appendix 5 Instruction manual Appendix 6 Casing Drg Impeller Drg Shaft Drg GA Drg Erection Drg
1.2 CONTROLS			
1.2.1 SAFETY AND RELIABILITY OF CONTROL SYSTEMS	YES	* Control panel supplier to confirm compliance with 1.2.1-1.2.8.	Appendix 6 Logic Description P&I Drg
1.2.2 CONTROL DEVICES	YES	See 1.2.1	
1.2.3 STARTING	YES	See 1.2.1	
1.2.4 STOPPING DEVICE	YES	See 1.2.1	
1.2.5 MODE SELECTION	YES	See 1.2.1	
1.2.6 FAILURE OF POWER SUPPLY	YES	See 1.2.1	

OPIC	APP	COMMENTS	FILE REF
.2.7 FAILURE OF HE CONTROL IRCUIT	YES	See 1.2.1	
.2.8 SOFTWARE	YES	See 1.2.1	
.3 PROTECTION GAINST MECHANICAL HAZARDS			
.3.1 STABILITY	YES	Consideration of stability has been included during design. Holding down points for baseplate are provided. Gearbox and engine bolted to baseplate. Pump is supported inside casson.	Appendix 6 GA Drg
.3.2 RISK OF BREAKUP DURING OPERATION	YES	Design calculations shown that pumpset design is sound if operated within stated limits.	Appendix 2 Pump design checklist Appendix 4 Pump data sheet
.3.3 RISKS DUE O FALLING OR JECTED OBJECTS	YES	Pump & motor casings will contain any fragments.	Appendix 6 Pump casing drawing Motor Casing Drg
.3.4 RISKS DUE O SURFACES, DGES, OR ANGLES	YES	Drawings of components with exposed surfaces show no risk.	Appendix 6 Section arrangement drawing GA Drg Pump casing drawing
.3.5 RISKS RELATED TO COMBINED MACHINERY	NO	Pump has only one defined operating mode.	
.3.6 RISKS RELATED TO VARIATIONS IN ROTATIONAL SPEED OF TOOLS	NO	Speed controlled by engine governor. Overspeed trip provided on engine.	Appendix 5 Instruction manual
.3.7 PREVENTION OF RISKS RELATED TO MOVING PARTS	YES	Wire mesh guard fitted over exposed shaft at coupling. and gland, coupling guard fitted between engine and gearbox.	Appendix 6 GA drawing Parts list

TOPIC	APP	COMMENTS	FILE REF
1.3.8 CHOICE OF PROTECTION AGAINST RISKS RELATED TO MOVING PARTS	YES	Pumpset is fitted with a fixed guard at the gland & coupling locations in compliance Section 1.4.2.1.	Appendix 6 GA drawing Parts list
1.4 REQUIRED CHARACTERISTICS OF GUARDS AND PROTECTION DEVICES			
1.4.1 GENERAL REQUIREMENT	YES	From experience the guards fitted at the pump gland and coupling locations comply with this section.	Appendix 6 GA drawing Parts list
1.4.2 SPECIAL REQUIREMENTS FOR GUARDS			
1.4.2.1 FIXED GUARDS	YES	Guard fitted at the gland and coupling locations with screws.	Appendix 6 Parts list
1.4.2.2 MOVABLE GUARDS	NO	No movable guards are fitted.	
1.4.2.3 ADJUSTABLE GUARDS RESTRICTING ACCESS	NO	No adjustable guards are fitted.	
1.4.3 SPECIAL REQUIREMENTS FOR PROTECTION DEVICES	NO	None fitted.	
1.5 PROTECTION AGAINST OTHER HAZARDS			
1.5.1 ELECTRICAL SUPPLY	YES	* Control panel supplier to confirm design code used. Other voltages present are not hazardous.	Appendix 2 Pump design checklist Appendix 6 P&I Drg
1.5.2 STATIC ELECTRICITY	YES	All components of the pumpset are earthed to the baseplate.	Appendix 6 GA Drg. & Parts list
1.5.3 ENERGY SUPPLY OTHER THAN ELECTRICITY	YES	* Engine supplier to confirm design code used.	

TOPIC	APP	COMMENTS	FILE REP
1.5.4 ERRORS OF FITTING	YES	From experience there is minimal risk.	Appendix 6 GA drawing Instruction Manual
1.5.5 EXTREME TEMPERATURES	NO	Pumpset is supplied without lagging as agreed in the contract. Customer must lag all exposed areas with a surface temperature in accordance with EN563.	
1.5.6 FIRE	NO	Pump is not considered a fire risk. * Engine supplier to confirm what measures are taken to prevent fire.	
1.5.7 EXPLOSION	NO	Pumpset is not suitable for use in an explosive environment.	
1.5.8 NOISE	YES	Reasonable consideration taken.	Appendix 4 Pump data sheet
1.5.9 VIBRATION	YES	Reasonable consideration taken. Impeller balance.	Appendix 3 Pump impeller Balance Certificate
1.5.10 RADIATION	NO	Pumpset operation does not involve any possible emission of radiation.	
1.5.11 EXTERNAL RADIATION	NO	Pumpset will be designed to unaffected by external radiation if specified by the customer at time of order. Otherwise standard environmental conditions apply.	Appendix 4 Pump data sheet
1.5.12 LASER EQUIPMENT	NO	No laser equipment is incorporated in the pump.	
1.5.13 EMISSION OF DUST, GASES, ETC	YES	Any leakage from pump seal can be drained using drain connection provided.	Appendix 4 Pump data sheet Appendix 6 GA drawing
1.6 MAINTENANCE			
1.6.1 MACHINERY MAINTENANCE	YES	Complies except for gland adjustment operation. * Instruction manual to detail safe adjustment of the gland while pump is running.	Appendix 5 Instruction manual Appendix 6 GA Drg

TOPIC	APP	COMMENTS	FILE REP
1.6.2 ACCESS TO OPERATING POSITION AND SERVICING POINTS	NO	Additional access not required.	Appendix 6 GA Drg
1.6.3 ISOLATION OF ENERGY SOURCES	YES	* Customer to provide means of isolating electrical supply. * Provision for isolation of main battery supply to be considered.	Appendix 6 P&I Drg
1.6.4 OPERATOR INTERVENTION	YES	Reduced as much as possible within the scope of pump design.	Appendix 6 GA drawing
1.6.5 CLEANING OF INTERNAL PARTS	YES	Internal parts can be cleaned by circulation of a cleaning medium through the pump	Appendix 6 Section Arrangement drawing Pump casing drawing
1.7 INDICATORS			
1.7.0 INFORMATION DEVICES	YES	* Control panel supplier to confirm compliance	Appendix 6 Logic Description P&I Drg
1.7.1 WARNING DEVICES	YES	* Control panel supplier to confirm compliance	Appendix 6 Logic Description P&I Drg
1.7.2 WARNING OF RESIDUAL RISKS	YES	* Warning of gland adjustment required in instructions and on machinery	Appendix 5 Instruction manual
1.7.3 MARKING	YES	* Nameplate to be reviewed to ensure all safety information has been included eg maximum speed suction pressure and temperature	
1.7.4 INSTRUCTIONS	YES	* Instructions to be checked for compliance	Appendix 5 Instruction manual

HAZARD IDENTIFICATION SHEET DATA

Study title: HAZID STUDY

System/Area: FIREPUMP

Date: 1.6.94

Participants
G FOY - CONTRACTS
G DONALDSON - C&I
P E McFADDEN - DO
F C PORTEOUS - WES
A GREIG - S&S

A FERGUSON - TEST
L MAXWELL - DESIGN

Ref No.	Generic hazard & guideword	Event description	Consequence/ escalation	Prevention/contr ol mitigating factor	Risk index Cons freq	Action required/ comment
1	Gas ingress	Gas in safe area	Gas in safe area	Leak tight seal/welds and cable glands	Low	Gas tight seals and welds (BRV/WPL) Cable glands (WPL)
2	Gas ingress	Room contaminated with exhaust fumes Danger to personnel	Room contaminated with exhaust fumes. Danger to personnel	Leak tight seals on exhaust pipework. Partially tested during string test.	Low	Gas tight seals on exhaust pipework (BRV)
3	Gas ingress	Room contaminated with explosive gas	Room contaminated with explosive gas	Insufficient quantity to produce danger	Nil	No action required. Handled by normal HVAC system
4	Gas ingress	Leakage from air start supply cylinders	No dangerous emissions. Low pressure in air start cylinders. Air start inoperative	Low pressure alarm. Switch set to alarm with sufficient air remaining in cylinders to start the engine.	Medium	Check system during routing maintenance (BRV)
5	Gas ingress	Gas in diesel engine air intake	Engine overspeed. High exhaust gas temperature. Engine damage	Overspeed protection.	Low	Air intake ducting to be routed from safe area (BRV)

HAZARD IDENTIFICATION SHEET DATA

Study title: HAZID STUDY

System/Area: FIREPUMP

Date: 1.6.94

Participants
G FOY - CONTRACTS
G DONALDSON - C&I
P E McFADDEN - DO
F C PORTEOUS - WES
A GREIG - S&S

A FERGUSON - TEST
L MAXWELL - DESIGN

Sheet 2 of 6

Ref No.	Generic hazard & guideword	Event description	Consequence/ escalation	Prevention/control mitigating factor	Risk index Cons freq	Action required/ comment
6	Gas ingress	Exhaust gas emissions setting off fire alarm systems	Emergency procedures instigated. Loss of production.	Exhaust pipe - work routed to an area away from alarms	Medium	Gas tight seals and welds (BRV/WPL) Cable glands (WPL)
7	Gas ingress	Gas leakage into enclosure	Pockets of gas under skid. Possible explosion.	Integrity of floor penetrations	Low	Gas tight seals on exhaust pipework (BRV)
8	Fire (jet)	Fuel line rupture	Jet fire	Heavy schedule fuel lines plus remotely operated fuel shut off valve	Medium	No action required. Handled by normal HVAC system
9	Fire (pool)	Leakage from fuel tank	Potential pool fire	Drain pan on tank sighted under potential leakage source piped away via baseplate drain pan to drainage area	Low	Check system during routing maintenance (BRV)
10	Fire (flash)	Failure of temperature control switch on gearbox lub oil heater	Flash fire	maximum temperature attainable is calculated at 65% of oil flash point. Heater would fail prior to the flash temperature being reached	Low	Choice of switch with higher cut off or a high temperature switch with alarm available. WPL happy with existing arrangement. BRV to action change if required

118

Study title: HAZID STUDY

System/Area: FIREPUMP

Date: 1.6.94

Participants
G FOY - CONTRACTS
G DONALDSON - C&I
P E McFADDEN - DO
F C PORTEOUS - WES
A GREIG - S&S

A FERGUSON - TEST
L MAXWELL - DESIGN

Sheet 3 of 6

Ref No.	Generic hazard & guideword	Event description	Consequence/ escalation	Prevention/control mitigating factor	Risk index Cons freq	Action required/ comment
11	Explosion (fire)	Crankcase emissions	Ignition of gas	Crankcase of breather vented to safe area	Low	BRV to vent through roof of enclosure and to consider spark arrestor WPL to show TP point (N) on general arrangement O & M manual to include maximum pressure within air start text (WPL)
12	Explosion (missiles)	Air start cylinders over pressurised	Cylinders bursting creating missiles	Warning name plate on air start equipment giving maximum charge pressure	Medium	
13	Pollution)	Failure of anti-fouling unit	Pump and piping fouled up or strainer blocked by marine growth	Visual indication of current failure	Low	No action required
14	Pollution	Engine air intake filter dirty/blocked	Loss of engine	Visual indicator on each filter with 'red line' warning	Low	No action required
15	Temperature	Room overheating after shut down (instruments and cables, batteries recommended max temp 50°C)	Eventual break-down of cables and shortened battery life	Reduce temperature	Medium	WPL to estimate room temp and effects on equipment. BRV to adequately rate the HVAC system

119

HAZARD IDENTIFICATION SHEET DATA

Study title: HAZID STUDY

B P ANDREW

Participants
G FOY - CONTRACTS
G DONALDSON - C&I
P E McFADDEN - DO
F C PORTEOUS - WES
A GREIG - S&S

A FERGUSON - TEST
L MAXWELL - DESIGN

System/Area: FIREPUMP

Date: 1.6.94

Ref No.	Generic hazard & guideword	Event description	Consequence/ escalation	Prevention/control mitigating factor	Risk index Cons freq	Action required/ comment
16	Temperature	Hot equipment during running and after shutdown	Injury to operating personnel	Crankcase of breather vented to safe area	Low	WPL to provide nameplates
17	Maintenance	Removing air start cylinders for recharging without closing isolating valves	Release of high pressure air. Potential injury to personnel	Warning name plate on air start equipment giving maximum charge pressure	Low	WPL to provide nameplates
18	Leakage	Joints on CW harness leakage	Loss of cooling failure of engine	Visual indication of current failure	Low	No action required
19	Leakage	Leakage of fuel oil'	Danger to personnel due to unsound surface. Failure of unit to operate	Visual indicator on each filter with 'red line' warning	Low	No action required
20	Leakage	Battery case damaged	Batteries enclosed within battery box	Reduce temperature	Low	Check battery condition during routine maintenance (BRV/BP)

120

Study title: HAZID STUDY

System/Area: FIREPUMP

Date: 1.6.94

Participants
G FOY - CONTRACTS
G DONALDSON - C&I
P E McFADDEN - DO
F C PORTEOUS - WES
A GREIG - S&S

A FERGUSON - TEST
L MAXWELL - DESIGN

Sheet 5 of 6

Ref No.	Generic hazard & guideword	Event description	Consequence/ escalation	Prevention/control mitigating factor	Risk index Cons freq	Action required/ comment
21	Location	Bel caisson and jacket caisson misalignment. Poor caisson support	Difficulty in installing pump. Movement of caisson during operation. Poor pump performance and reliability	Lead in bellmouth on caisson	Medium	WPL advise misalignment tolerances on caisson. BRV to advise caisson movements including 2g acceleration forces
22	Location	Sparks in exhaust gas	Ignition of external gas	Spark arrestor in silencer	Low	No action required
23	Location	Fuel shut off. Valve cable route produces exceptionally long cable	Cable stretches when used. Shut off valve fails to operate	Shorter cable route	Medium	WPL to advise maximum cable length
24	Location	Emergency escape from control panel region. Control panel do	Partial blockage of escape route	Location of fuel lines relative to control panel. Heavy schedule fuel lines	Low	Removal of 90° stop on control panel door
25	Location	Flames entering fuel tank via vent pipe	Escalation of fire. Fire pump inoperative	Deluge system. Fitting of flame arrestor into vent pipe	Low	BRV to consider the need for fitting flame arrestor
26	Control mechanisms	Air start button masked from view	Operator unable to locate start button. Unit fails to start	Location detailed in engine manual	Medium	WPL to provide directional nameplate near to location, clearly sited

121

HAZARD IDENTIFICATION SHEET DATA

Study title: HAZID STUDY

System/Area: FIREPUMP

Date: 1.6.94

Participants
G FOY - CONTRACTS
G DONALDSON - C&I
P E McFADDEN - DO
F C PORTEOUS - WES
A GREIG - S&S

A FERGUSON - TEST
L MAXWELL - DESIGN

Ref No.	Generic hazard & guideword	Event description	Consequence/ escalation	Prevention/contr ol mitigating factor	Risk index Cons freq	Action required/ comment
27	Human factor	Manual operation selected	No automatic start	Setting indicated on control panel and main control	Low	No action required
28	Human factor	Isolation valves not opened after re-charging cylinders	No air flow to start motor. Unit fails to start	Low pressure switch alarm	Medium	Review air start system and remove any unnecessary isolation valves
29	Human factor	Contact with rotating components	Injury to operator	Guard on all exposed rotating parts	Low	No action required
30	Human factor	Remote fuel shut off valve closed outwith emergency condition. Not reset	No fuel to engine. Unit fails to start or operates for short time	Failed to start signal from control panel. Second unit start initiated	Medium	BRV to locate remote fuel shut off valve mechanism within a lockable cabinet with breakable glass panel
31	Human factor	Equipment lifted using incorrect methods	Damage to equipment Potential injury to personnel	Installation and erection drawings. O & M manual gives itemised weight schedule	Low	No action required

WEIR PUMPS LIMITED

SAFETY ACTION SHEET - MAIN HAZID

CONTRACT NO:	13168
DESCRIPTION:	B.P. ANDREW

AREA OR SYSTEM:	FIREWATER PUMP

SAFETY ACTION (Refer H.I.D.S.)

REF NO: H.I.D.S. REF NO. 15

BRIEF DESCRIPTION: Refer to attached sheet and appropriate "Ref No." for action details.

INITIATOR: L MAXWELL   SIGNED: *L Maxwell*   DATE: 10/6/94

ACTION IMPLEMENTED (TO BE COMPLETED BY RECIPIENT)

FAX TO BRU/BP DATED 29/8/94 GIVING DETAILS OF ROOM TEMPERATURES AND INDICATING ANY EFFECTS WILL BE NEGLIGIBLE.

L MAXWELL
NAME/POSITION: DESIGN ENG SIGNED: *L Maxwell* DATE: 29/8/94

ACTION COMPLETE (TO BE COMPLETED BY INITIATOR)

INITIATOR: *L Maxwell*   DATE: 29/8/94

NOTE: On completion of "Action Implemented Section" sheet to be returned to initator

WEIR PUMPS LIMITED

SAFETY ACTION SHEET - MAIN HAZID

CONTRACT NO:	13168
DESCRIPTION:	B.P. ANDREW

AREA OR SYSTEM:	FIREWATER PUMP

SAFETY ACTION (Refer H.I.D.S.)

REF NO: H.I.D.S. REF NO. 16

BRIEF DESCRIPTION:　　Refer to attached sheet and appropriate "Ref No." for action
　　　　　　　　　　　details.

INITIATOR:　L MAXWELL　　　　SIGNED: *L Maxwell*　　DATE: 10/6/94

ACTION IMPLEMENTED (TO BE COMPLETED BY RECIPIENT)

INSTRUCTION ISSUED TO DRAWING OFFICE TO PROVIDE WARNING NAMEPLATES ON BOTH SIDES OF SKID TO HIGHLIGHT "HIGH TEMPERATURE".

L MAXWELL

NAME/POSITION: DESIGN ENG　SIGNED: *L Maxwell*　DATE: 13/6/94

ACTION COMPLETE (TO BE COMPLETED BY INITIATOR)

INSTRUCTIONS ISSUE BY D.O. FOR MANUFACTURE.

INITIATOR: *L Maxwell*　　　　DATE:　27/6/94

NOTE:　On completion of "Action Implemented Section" sheet to be returned to initator

WEIR PUMPS LIMITED

SAFETY ACTION SHEET - MAIN HAZID

CONTRACT NO:	13168
DESCRIPTION:	B.P. ANDREW

AREA OR SYSTEM:	FIREWATER PUMP

SAFETY ACTION (Refer H.I.D.S.)

REF NO: H.I.D.S. REF NO. 17

BRIEF DESCRIPTION: Refer to attached sheet and appropriate "Ref No." for action details.

INITIATOR: L MAXWELL   SIGNED: *L Maxwell*   DATE: 10/6/94

ACTION IMPLEMENTED (TO BE COMPLETED BY RECIPIENT)

INSTRUCTIONS ISSUED TO DRAWING OFFICE TO PROVIDE NAME PLATE ON AIR START EQUIP.T TO INSTRUCT USER TO VENT CYLINDER AND CLOSE ISOLATION VALUE PRIOR TO REMOVAL BEFORE RECHARGING.

L MAXWELL

NAME/POSITION: DESIGN ENG. SIGNED: *L Maxwell* DATE: 13/6/94

ACTION COMPLETE (TO BE COMPLETED BY INITIATOR)

INSTRUCTIONS ISSUED BY D.O. FOR MANUFACTURE

INITIATOR: *L Maxwell*   DATE: 27/6/94

NOTE: On completion of "Action Implemented Section" sheet to be returned to initator

WEIR PUMPS LIMITED

SAFETY ACTION SHEET - MAIN HAZID

CONTRACT NO:	13168
DESCRIPTION:	B.P. ANDREW

AREA OR SYSTEM:	FIREWATER PUMP

SAFETY ACTION (Refer H.I.D.S.)

REF NO: H.I.D.S. REF NO. 20

BRIEF DESCRIPTION: Refer to attached sheet and appropriate "Ref No." for action
details.

INITIATOR: L MAXWELL   SIGNED: *LMaxwell*   DATE: 10/6/94

ACTION IMPLEMENTED (TO BE COMPLETED BY RECIPIENT)

FAX SENT TO BRV/BP DATED 29/8/94
HIGHLIGHTING ACTION.

L. MAXWELL

NAME/POSITION: DESIGN ENG. SIGNED: *LMaxwell* DATE: 29/8/94

ACTION COMPLETE (TO BE COMPLETED BY INITIATOR)

INITIATOR: *LMaxwell*   DATE: 29/8/94

NOTE: On completion of "Action Implemented Section" sheet to be returned to initator

WEIR PUMPS LIMITED

SAFETY ACTION SHEET - MAIN HAZID

CONTRACT NO:	13168
DESCRIPTION:	B.P. ANDREW

AREA OR SYSTEM:	FIREWATER PUMP

SAFETY ACTION (Refer H.I.D.S.)

REF NO: H.I.D.S. REF NO. 21

BRIEF DESCRIPTION: Refer to attached sheet and appropriate "Ref No." for action
details.

INITIATOR: L MAXWELL   SIGNED: *L I Maxwell*   DATE: 10/6/94

ACTION IMPLEMENTED (TO BE COMPLETED BY RECIPIENT)

THIS ACTION HAS BEEN THE SUBJECT OF MUCH
DISCUSSION BETWEEN WPL AND BP DEV^L AND IS
DOCUMENTED ON VARIOUS FAXES. A COMPROMISED
SITUATION HAS BEEN REACHED WHICH IS ACCEPTABLE TO
BOTH. FAXES FILED IN RELEVANT JOB PACKET.
NAME/POSITION: L MAXWELL DESIGN ENG SIGNED: *L I Maxwell* DATE: 16/6/94

ACTION COMPLETE (TO BE COMPLETED BY INITIATOR)

INITIATOR: *L I Maxwell*   DATE: 16/6/94

NOTE: On completion of "Action Implemented Section" sheet to be returned to initator

WEIR PUMPS LIMITED

SAFETY ACTION SHEET - MAIN HAZID

CONTRACT NO:	13168
DESCRIPTION:	B.P. ANDREW

AREA OR SYSTEM:	FIREWATER PUMP

SAFETY ACTION (Refer H.I.D.S.)

REF NO: H.I.D.S. REF NO. 23

BRIEF DESCRIPTION: Refer to attached sheet and appropriate "Ref No." for action
details.

INITIATOR: L MAXWELL   SIGNED: *L Maxwell*   DATE: 10/6/94

ACTION IMPLEMENTED (TO BE COMPLETED BY RECIPIENT)

FAX SENT TO BRV/BP DATED 29/8/94 HIGHLIGHTING ACTION.

L MAXWELL
NAME/POSITION: DESIGN ENG SIGNED: *L Maxwell* DATE: 30/8/94

ACTION COMPLETE (TO BE COMPLETED BY INITIATOR)

INITIATOR: *L Maxwell*   DATE: 30/8/94

NOTE: On completion of "Action Implemented Section" sheet to be returned to initator

128

WEIR PUMPS LIMITED

SAFETY ACTION SHEET - MAIN HAZID

| CONTRACT NO: | 13168 |
| DESCRIPTION: | B.P. ANDREW |

| AREA OR SYSTEM: | FIREWATER PUMP |

SAFETY ACTION (Refer H.I.D.S.)

REF NO: H.I.D.S. REF NO. 24

BRIEF DESCRIPTION: Refer to attached sheet and appropriate "Ref No." for action
details.

INITIATOR: L MAXWELL   SIGNED: *L/Maxwell*   DATE: 10/6/94

ACTION IMPLEMENTED (TO BE COMPLETED BY RECIPIENT)

INSTRUCTION ISSUED TO ELECTRICAL FOREMAN
TO HAVE STOP REMOVED PRIOR TO UNIT
LEAVING WPL WORKS.

L MAXWELL

NAME/POSITION: DESIGN ENG .SIGNED: *L/Maxwell* DATE: 11/8/94

ACTION COMPLETE (TO BE COMPLETED BY INITIATOR)

INITIATOR: *L/Maxwell*   DATE: 11/8/94

NOTE: On completion of "Action Implemented Section" sheet to be returned to initator

FIRE WATER PUMP

SAFETY ACTION SHEET LOG   MAIN HAZOP

Action No	Date	Subject	Action
627	10 May 94	Main valve to diesel engine closed	Completed
628	10 May 94	Valve on vent left closed	Completed
629	10 May 94	Flexible connection missing between secondary air start system and the diesel engine	Completed
630	10 May 94	Block valve up stream and down stream of PVC 36004 closed	Completed
631	10 May 94	Block valve up stream and down stream of PVC 36009 closed	Completed
632	10 May 94	Block valve up stream and down stream of PVC 36008 closed	Completed
633	10 May 94	Burst tube on lube oil cooler	Pending
634	10 May 94	Consider installing PIs downstream of PCVs 36004 and 36008	Pending

Rev 0   Issued 2 June 94

HAZARD AND OPERABILITY STUDY ACTION SHEET
DATA FILE

ACTION ON: Weir – MECH	RESPOND BY:

ACTION NO: 627	MEETING DATES: 14/2/94

DOCUMENT REFERENCE: PME06XCO501+PID-1011-001/2/D2   REVISION: 2
TITLE: FIREWATER PUMPS

PLANT SECTION: (HAZOP Table 631/2)
SECONDARY AIR START TO THE DIESEL ENGINE

CAUSE:
Main valve to diesel engine closed

CONSEQUENCE:
Unable to crank machine in an emergency

SAFEGUARDS/REMARKS:
None

ACTION:
Consider locking open main valve

RESPONSE: DATED: 1/6/94

THE VALVE IN QUESTION IS IN CLOSE PROXIMITY TO THE AIR START BUTTON AND IS EASILY ACCESSIBLE.

IF IT WERE TO BE LEFT CLOSED THERE WOULD BE NO PROBLEM IN CORRECTING THE SITUATION AND THE TIME TAKEN NEGLIGIBLE.

A NAMEPLATE ADDED TO AIRSTART SYSTEM HIGH-LIGHTING VALVE SHOULD BE LEFT OPEN.

OPERATING AND MAINTENANCE MANUAL WILL INCLUDE TEXT WHICH HIGHLIGHTS THE IMPORTANCE OF VALVE BEING OPEN AND TO CHECK VALVE IF AIR START MOTOR DOES NOT OPERATE WHEN BUTTON IS PUSHED. THE DANGER IS A LOCKABLE VALVE MAY BE LOCKED CLOSED IN ERROR THUS ISOLATING THE AIR SUPPLY. IT IS DECIDED THEREFORE MAIN VALVE TO REMAIN UNLOCKABLE.   SIGNED: L/Maxwell

ENTER YOUR RESPONSE IN THE BOX ABOVE THEN SIGN AND RETURN THIS FORM TO:
B Brazier

NOTES: (For use of HAZOP Secretary only)

SAFETY ACTION SHEET - MAIN HAZOP

ACTION ARISING FROM ADP-SA-REP-0552-000	ACTION NO: SAS-MHA-627

AREA OR SYSTEM: See attached sheet	Date: See attached sheet

SAFETY ACTION: (description from original report)

ACTN NO. 627   SEE ATTACHED SHEET

Signed: *B'W. Brazier*   Name: *B.BRAZIER*   Date: 4/5/94

ACTION TAKEN TO RESOLVE: PLEASE ENTER YOUR RESPONSE ON THIS SHEET

See attached

action for implementation.
check if nameplate added.

Signed: Name: Date:

This Action has been Approved:

Project Engineer: *ABR*   Date: 16/9/94
Lead Engineer: Date:
Safety Engineer: Date: 16/9/94

This Action has been implemented: Cost/Savings:

Safety Engineer: Date: £:

HAZARD AND OPERABILITY STUDY ACTION SHEET
DATA FILE

ACTION ON: Weir	RESPOND BY:

ACTION NO: 628	MEETING DATES: 14/2/94

DOCUMENT REFERENCE: PME06XC0501+PID-1011-001/2/D2   REVISION: 2
TITLE: FIREWATER PUMPS

PLANT SECTION: (HAZOP Table 631/2)
SECONDARY AIR START TO THE DIESEL ENGINE

CAUSE:
Valve on vent left open

CONSEQUENCE:
Loss of air to diesel engine when required to start

SAFEGUARDS/REMARKS:
None

ACTION:
Consider locking vent valve closed

RESPONSE: DATED: 1/6/94
LOW PRESSURE ALARM INCORPORATED IN
AIR START SYSTEM TO FLAG UP LOW PRESSURE.

AIR ESCAPING FROM VENT VALVE WOULD BE AUDIBLE.

NAMEPLATE ADDED TO AIRSTART SYSTEM WARNING
AGAINST LEAVING VENT VALVE OPEN.

IT IS CONSIDERED UNNECESSARY TO AUGMENT
THE ABOVE FEATURES WITH LOCKABLE VALVES.

SIGNED: L I Maxwell

ENTER YOUR RESPONSE IN THE BOX ABOVE THEN SIGN AND RETURN THIS FORM TO:
B Brazier

NOTES: (For use of HAZOP Secretary only)

SAFETY ACTION SHEET - MAIN HAZOP

ACTION ARISING FROM ADP-SA-REP-0552-000	ACTION NO: SAS-MHA-628

AREA OR SYSTEM: See attached sheet	Date: See attached sheet

SAFETY ACTION: (description from original report)

ACTN NO. 628   SEE ATTACHED SHEET

Signed: B.W. Brazier   Name: B. BRAZIER   Date: 4/5/94

ACTION TAKEN TO RESOLVE: PLEASE ENTER YOUR RESPONSE ON THIS SHEET

See attached

action for implementation.
check if nameplate added.

Signed: Name: Date:

This Action has been Approved:
Project Engineer: ABR   Date: 16/9/94
Lead Engineer: Date:
Safety Engineer: KKarry   Date: 16/9/94

This Action has been implemented: Cost/Savings:
Safety Engineer: Date: £:

HAZARD AND OPERABILITY STUDY ACTION SHEET
DATA FILE

ACTION ON: Weir ~ MECH		RESPOND BY:
ACTION NO: 629	MEETING DATES: 14/2/94	

DOCUMENT REFERENCE: PME06XC0501+PID-1011-001/2/D2   REVISION: 2
TITLE: FIREWATER PUMPS

PLANT SECTION: (HAZOP Table 631/2)
SECONDARY AIR START TO THE DIESEL ENGINE

ACTION:
Show flexible connection between secondary air start system and diesel engine

RESPONSE: DATED: 1/6/94

FLEXIBLE CONNECTION SHOWN ON
P & I DIAGRAM N° AO-501045.

SIGNED: L Maxwell

ENTER YOUR RESPONSE IN THE BOX ABOVE THEN SIGN AND RETURN THIS FORM TO:
B Brazier

NOTES: (For use of HAZOP Secretary only)

SAFETY ACTION SHEET - MAIN HAZOP

ACTION ARISING FROM ADP-SA-REP-0552-000	ACTION NO: SAS-MHA-629

AREA OR SYSTEM: See attached sheet	Date: See attached sheet

SAFETY ACTION: (description from original report)

ACTN NO. 629   SEE ATTACHED SHEET

Signed: B.W.Brazier   Name: B.BRAZIER   Date: 4/5/94

ACTION TAKEN TO RESOLVE: PLEASE ENTER YOUR RESPONSE ON THIS SHEET

see attached

Signed: Name: Date:

This Action has been Approved:

Project Engineer: ABR   Date: 16/9/94
Lead Engineer: Date:
Safety Engineer: Date: 16/9/94

This Action has been implemented: Cost/Savings:
Safety Engineer: Date: 16/9/94   £:

HAZARD AND OPERABILITY STUDY ACTION SHEET
DATA FILE

ACTION ON: Weir – MECH	RESPOND BY:

ACTION NO: 630	MEETING DATES: 14/2/94

DOCUMENT REFERENCE: PME06XC0501+PID-1011-001/2/D2   REVISION: 2
TITLE: FIREWATER PUMPS

PLANT SECTION: (HAZOP Table 631/3)
SEAWATER COOLING FROM PUMP DISCHARGE TO LUBE OIL COOLER, ENGINE COOLER
AND AIR COOLER

CAUSE:
Block valves upstream and downstream of PCV36004 closed

CONSEQUENCE:
High gearbox temperature and potential mechanical damage

SAFEGUARDS/REMARKS:
This is likely to be undetected

ACTION:
Consider locking open block valves

RESPONSE: DATED: 1/6/94

LOCKABLE VALVES INCORPORATED IN GEARBOX
COOLING HARNESS. LOCKED OPEN POSITION
INDICATED ON P&I DIAGRAM Nº AO-50104S.
OPERATING AND MAINTENANCE MANUAL WILL
INCLUDE NECESSARY INSTRUCTIONS WITHIN
ITS TEXT.

SIGNED: L J Maxwell

ENTER YOUR RESPONSE IN THE BOX ABOVE THEN SIGN AND RETURN THIS FORM TO:
B Brazier

NOTES: (For use of HAZOP Secretary only)

SAFETY ACTION SHEET - MAIN HAZOP

ACTION ARISING FROM ADP-SA-REP-0552-000	ACTION NO: SAS-MHA-630

AREA OR SYSTEM: See attached sheet	Date: See attached sheet

SAFETY ACTION: (description from original report)

ACTN NO. 630   SEE ATTACHED SHEET

Signed: *B.W. Brazier*   Name: *B BRAZIER*   Date: 4/5/94

ACTION TAKEN TO RESOLVE: PLEASE ENTER YOUR RESPONSE ON THIS SHEET

See attached
_____ '' _____

action for implementation.
Check lockable valves included
Ensure lockable valves shown on
                 P&ID

Signed: Name: Date:

This Action has been Approved:
Project Engineer: Date: 16/9/94
Lead Engineer: Date:
Safety Engineer: Date: 16/9/94

This Action has been implemented: Cost/Savings:
Safety Engineer: Date: £:

138

HAZARD AND OPERABILITY STUDY ACTION SHEET
DATA FILE

ACTION ON: Weir — Mech	RESPOND BY:

ACTION NO: 631	MEETING DATES: 14/2/94

DOCUMENT REFERENCE: PME06XCO501+PID-1011-001/2/D2   REVISION: 2
TITLE: FIREWATER PUMPS

PLANT SECTION: (HAZOP Table 631/3)
SEAWATER COOLING FROM PUMP DISCHARGE TO LUBE OIL COOLER, ENGINE COOLER
AND AIR COOLER

CAUSE:
Block valves upstream and downstream of PCV36009 closed

CONSEQUENCE:
Increase in temperature of diesel engine and potential mechanical damage

SAFEGUARDS/REMARKS:
None

ACTION:
Consider locking open these block valves

RESPONSE: DATED: 1/6/94

LOCKABLE VALVES INCORPORATED IN COOLING
WATER HARNESS. LOCKED OPEN POSITION
INDICATED ON P&I DIAGRAM N° AO-501045.
OPERATING AND MAINTENANCE MANUAL WILL
INCLUDE NECESSARY INSTRUCTIONS WITHIN
ITS TEXT

SIGNED: LI Maxwell

ENTER YOUR RESPONSE IN THE BOX ABOVE THEN SIGN AND RETURN THIS FORM TO:
B Brazier

NOTES: (For use of HAZOP Secretary only)

SAFETY ACTION SHEET - MAIN HAZOP

ACTION ARISING FROM ADP-SA-REP-0552-000	ACTION NO: SAS-MHA-631

AREA OR SYSTEM: See attached sheet	Date: See attached sheet

SAFETY ACTION: (description from original report)
ACTN NO. 631   SEE ATTACHED SHEET
Signed: *B.W. Brazier*   Name: B. BRAZIER   Date: 4/5/94

ACTION TAKEN TO RESOLVE: PLEASE ENTER YOUR RESPONSE ON THIS SHEET

See attached

Action for implementation.
check lockable valves included
Ensure   "   "   shown on P+ID

Signed: Name: Date:

This Action has been Approved:
Project Engineer: PBR   Date: 16/9/94
Lead Engineer: Date:
Safety Engineer: Date: 16/9/94

This Action has been implemented: Cost/Savings:
Safety Engineer: Date: £:

HAZARD AND OPERABILITY STUDY ACTION SHEET
DATA FILE

ACTION ON: Weir — MECH	RESPOND BY:

ACTION NO: 632	MEETING DATES: 14/2/94

DOCUMENT REFERENCE: PME06XC0501+PID-1011-001/2/D2   REVISION: 2
TITLE: FIREWATER PUMPS

PLANT SECTION: (HAZOP Table 631/3)
SECONDARY COOLING FROM PUMP DISCHARGE TO LUBE OIL COOLER, ENGINE COOLER AND AIR COOLER

CAUSE:
Block valve downstream of PCV36008 closed

CONSEQUENCE:
No air cooling in enclosure when required during a fire

SAFEGUARDS/REMARKS:
None

ACTION:
Consider locking open this block valve

RESPONSE: DATED: 1/6/94

LOCKABLE VALVES INCORPORATED IN COOLING WATER HARNESS. LOCKED OPEN POSITION INDICATED ON P & I DIAGRAM N° AO-501045.
OPERATING AND MAINTENANCE MANUAL WILL INCLUDE NECESSARY INSTRUCTIONS WITHIN ITS TEXT.

SIGNED: L/Maxwell

ENTER YOUR RESPONSE IN THE BOX ABOVE THEN SIGN AND RETURN THIS FORM TO:
B Brazier

NOTES: (For use of HAZOP Secretary only)

141

<u>SAFETY ACTION SHEET - MAIN HAZOP</u>

ACTION ARISING FROM ADP-SA-REP-0552-000	ACTION NO: SAS-MHA-632

AREA OR SYSTEM: See attached sheet	Date: See attached sheet

SAFETY ACTION: (description from original report)
ACTN NO. 632   SEE ATTACHED SHEET
Signed: B.W. Brazier   Name: B. BRAZIER   Date: 4/5/94

ACTION TAKEN TO RESOLVE: PLEASE ENTER YOUR RESPONSE ON THIS SHEET
See attached action for implementation. check lockable valves included Ensure   "   "   are on P+ID.
Signed: Name: Date:

This Action has been Approved:	
Project Engineer: ABR	Date: 16/9/94
Lead Engineer:	Date:
Safety Engineer:	Date: 16/9/94

This Action has been implemented:	Cost/Savings:
Safety Engineer: Date:	£:

142

SAFETY ACTION SHEET - MAIN HAZOP

ACTION ARISING FROM ADP-SA-REP-0552-000	ACTION NO: SAS-MHA-633

AREA OR SYSTEM: See attached sheet	Date: See attached sheet

SAFETY ACTION: (description from original report)

ACTN NO. 633   SEE ATTACHED SHEET

Signed: B.W.Brazier   Name: B.BRAZIER   Date: 4/5/94

ACTION TAKEN TO RESOLVE: PLEASE ENTER YOUR RESPONSE ON THIS SHEET

Signed: Name: Date:

This Action has been Approved:

Project Engineer: ABR   Date: 16/9/94
Lead Engineer: Date:
Safety Engineer: KLGray   Date: 16/9/94

This Action has been implemented: Cost/Savings:

Safety Engineer: Date: 16/9/94   £:

143

INDEX

S

T

U

V

W